T0291774

Healthcare Strategies and Planning for Social Inclusion and Development

Healthcare Strategies and Planning for Social Inclusion and Development

Volume 2: Social, Economic, and Health Disparities of Rural Women

Basanta Kumara Behera
Former Director, Advanced Centre for Biotechnology, Rohtak, Haryana, India

Ram Prasad
Associate Professor, Department of Botany, Mahatma Gandhi Central University, Motihari, Bihar, India

Shyambhavee Behera
Department of Community Medicine, University College of Medical Sciences, New Delhi, Delhi, India

Academic Press is an imprint of Elsevier
125 London Wall, London EC2Y 5AS, United Kingdom
525 B Street, Suite 1650, San Diego, CA 92101, United States
50 Hampshire Street, 5th Floor, Cambridge, MA 02139, United States
The Boulevard, Langford Lane, Kidlington, Oxford OX5 1GB, United Kingdom

ISBN 978-0-323-90447-6

For information on all Academic Press publications
visit our website at https://www.elsevier.com/books-and-journals

Publisher: Andre G. Wolff
Acquisitions Editor: Elizabeth Brown
Editorial Project Manager: Sam W. Young
Production Project Manager: Swapna Srinivasan
Cover Designer: Matthew Limbert

Typeset by STRAIVE, India

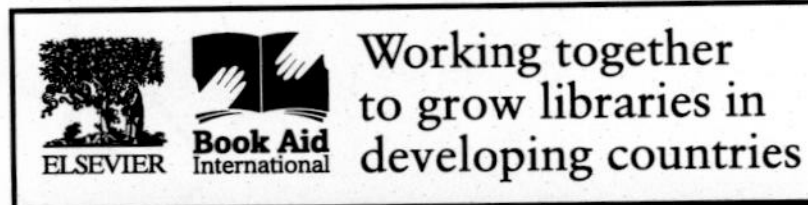

Contents

About the authors

Dr. Basanta Kumara Behera was a professor of biotechnology at three distinguished Indian universities, where he had been regularly associated with teaching and research at postgraduate level courses on topics related to medical biotechnology, biopharmaceutical, microbial process development, drug designing, bioenergy management, and biomass processing technology since 1978. In 2009 he joined an MNC company as an adviser for specialty chemicals production and drug design through microbial process technology. Prof. Behera is associated with companies of national and international repute as a technical adviser for the production of biopharmaceuticals under CGMP norms. Dr. Behera enjoys the credit for having authored various books published or in progress by CRC Press, USA; Springer-Verlag, Germany; Elsevier Inc. Cambridge, USA; CABI, Nosworthy Way, Wallingford, Oxfordshire.

Dr. Ram Prasad is Assistant Professor at the Amity Institute of Microbial Technology, Amity University, Uttar Pradesh, India. His research interest includes plant-microbe interactions, sustainable agriculture, and microbial nanobiotechnology. Dr. Prasad has more than a hundred publications to his credit (including research papers and book chapters and five patents issued or pending) and has edited or authored several books. He has 11 years of teaching experience and has been awarded the Young Scientist Award (2007) and Prof. J.S. Datta Munshi Gold Medal (2009) by the International Society for Ecological Communications; FSAB fellowship (2010) by the Society for Applied Biotechnology; Outstanding Scientist Award (2015) in the field of microbiology by Venus International Foundation; and the American Cancer Society UICC International Fellowship for Beginning Investigators (USA, 2014). In 2014–2015, Dr. Prasad served as Visiting Assistant Professor in the Department of Mechanical Engineering at Johns Hopkins University, USA.

Dr. Shyambhavee Behera is a medical graduate from Lady Hardinge Medical College, New Delhi, and holds an MD degree from University College of Medical Sciences, New Delhi. She has been working in the field of noncommunicable disease, immunization, maternal and child health, epidemiology, and health administration to upgrade and bring amendments in community for sustainable life pattern with good health. Currently, she working as Senior Resident, Department of Community Medicine, University College of Medical Sciences, New Delhi, India. As a coauthor, she has published a book titled *Move Towards Zero Hunger* with Springer Nature, Singapore Pte. Ltd. Dr. Shyambhavee is also the coauthor of the book series "New Paradigms of Living Systems" (Springer Nature, Singapore Pte. Ltd.)

Preface

This piece of work presents the paradigm of women's life in rural area and their social significance in developing a family, community and countryside. It also highlights how rural women and girls face arrays of diversified problem like cheap daily labor to manage livelihood; managing agricultural activities; taking care of children and old aged family members, even in health crisis. So, the book is designed to convince the readers to find way-out, and challenge in bringing social remediation to relief rural women from social exclusion and minimize their free labor activities and bitter confrontation with the parochial and deep rooted prejudice of traditional social norms.

The chapter begins with the concept of rural locality which is defined in various forms on the basis of geographical location and demographic structure. It also explains how the women play a key role in developing structure and function of rural community with special reference to agricultural productivity. In spite of so much efforts and scarifies, the rural women, still, confront lot of hurdles to avail the access for better health, education and sustainable lively hood. In this connection various social determinants like gender inequality, starvation, non availability of basic nutrients etc. are described in detailed, on the basis of social exclusion, disparity, aging issues, domestic violence, and health problem like obstetric and reproduction.

The second chapter covers how rural women confront with poorer health outcomes due to inadequate access to health care facilities, non availability of well trained healthcare work force. This chapter also highlights how women in rural areas face constraints in engaging in economic activities because of gender based discrimination and social norms, disproportionate. The last part of this section explains how most of the rural areas of developing countries are lacking of proper access to education facilities, including facilities for internet availability.

The subsequent chapter explains life cycle vulnerabilities of rural women from "womb to tomb." It also narrates how women are to be respected for the right to live from violence and supposed to free from social violence and discrimination, to enjoy the highest attainable standard of physical and mental health, to be educated, to own property, to vote, and to earn an equal wage.

Chapter 4 begins with the biological significance of quality food for a sustainable healthy life, especially reference to rural women residing in cluster of village life in a scatted form. Before getting into the depth of the subject maximum attention is given to define health and nutrition in simple manner to make the clear how quality nutrient is helpful for infants, young children and elderly person. In this connection it has been explain how to first understand the population health status through various biological indicators, and to take care of health, accordingly. It has been explained how various international agencies have been sincerely involved in initiating health risks handling project, world widely. In order to meet the high expensive budget to access health care facilities, how various types of health insurance coverage have been under operation condition in different countries is also explained.

The last chapter narrates the social significance of women's right to access equality, as the men enjoy in the society. In this connection it has been explained how inequality between genders not only affect individual but change entire economy scenario, both at national and global level. So, it has been well explained how to bring awareness on women's right by campaigning celebration of International Day for Women. Current, issues on gender equality under extreme socio-economic problem rose due to COVID-19 pandemic and lockdown of school, institutes and other offices, and supplementary unpaid work load on rural women, while managing family and agricultural services have been well presented with facts and figures. Additionally, the health problem being faced by women, working on the frontline has also well documented with global data survey reports from reliable international healthcare agencies like WHO, UN Women Organization, and UNICEF. It has also informed how safety and security measures of women working in pandemic effected work places are being taken by these organizations. The especial attraction of this chapter is on the various issues on unpaid women's labor, and how to reduce the work load of rural women through various labor saving technologies, easily acceptable to rural women has been presented with factual data and figures.

As this piece of work excessively covers the depth and breadth on the latest scenario of arrays of diversified life of women of rural areas and the possible remediation of their dismay life due to poverty, hunger and social discrepancy and social exclusion, the authors recommend this book both for basic scientists and seasoned researchers in the field of medical sciences, social sciences, community health management, and policy makers.

The authors are extremely thankful to Mrs. Asha Sharma, Katyayanee and Gopal for their help in computer graphics design, model preparations, and data compilation while preparing and giving shape to the manuscript. The kind cooperation of the publisher is highly acknowledged.

Abbreviations

ACS	American Community Survey
ANC	antenatal care
AP	adolescent pregnancy
ASHA	Association for Sanitation and Health Activities
BMI	body mass index
BPL	below poverty line
CMIE	Centre for Monitoring the Indian Economy
CSOs	Civil Society Organizations
EmOC	emergency obstetric care
EPMM	ending preventable maternal mortality
EU	European Union
FAO	Food Agriculture Organization
FBV	Epstein Barr virus
FDA	Food and Drug Administration
FFDCA	Federal Food, Drug, and Cosmetic Act
FGM	female genital mutilation
FGM/C	female genital mutilation/cutting
GNMF	Global Nutrition Monitoring Framework
HbA1c	glycosylated hemoglobin
HepB	hepatitis B
Hib	*Haemophilus influenzae* type b
HIV	human immunodeficiency virus
ICPD	International Conference on Population and Development
ID	iron deficiency
IDD	iodine deficiency
IFAD	International Fund for Agricultural Development
ILO	International Labour Organization
IMF	International Monetary Fund
IPU	Inter-Parliamentary Union
ITCILO	International Training Centre of the International Labour Organization
ITU	International Telecommunication Union
IWD	International Women's Day
LBW	low birth weight
MDGs	millennium development goals
MMR	maternal mortality rate
MoHFW	Ministry of Health and Family Welfare
MWCD	Ministry of Women and Child Development
NAA	National Assessment of Adult Literacy
NCDs	noncommunicable diseases
NGO	Nongovernmental Organization
NHI	National Health Insurance

NPOP	National Policy on Older Person
NSS	National Sample Survey
NSSO	National Sample Survey Organization
NWD	National Women's Day
PMJAY	Pradhan Mantri Jan Arogya Yojana
PMM	preventable maternal mortality
PNM	perinatal mortality
PPE	personal protective equipment
RIDP	Rural Income Diversification Project
RSBY	Rashtriya Swaathya Bima Yojana
RV	Rotavirus
SAGE	Study on Global Aging and Adult Health
SDGs	sustainable development goals
SDH	social determinants of health
SHIS	statutory health insurance system
SIDA	Swedish International Development Cooperation Agency
SIDS	sudden infant death syndrome
SMM	severe maternal morbidity
SNNPR	Southern Nations and Nationalities of Peoples' Region
SRHR	sexual and reproductive health and rights
SSA	sub-Saharan Africa
STIs	sexually transmitted infections
UHC	universal health coverage
UK	United Kingdom
UN	United Nations
UNCTAD	United Nations Conference on Trade and Development
UNEP	United Nations Environment Programme
UNESCO	United Nations Educational, Scientific and Cultural Organization
UNFPA	United Nations Fund for Population Activities
UNFPS	United Nations Population Fund
UNICEF	United Nations International Children's Emergency Fund
VAD	vitamin A deficiency
WB	World Bank
WFP	World Food Programme
WHA	World Health Assembly
WHO	World Health Organization

CHAPTER

Rural women's health disparities

1

1.1 What is rural area?

Generally, rural area or countryside is a geographic area that is located outside towns and cities. Cities, towns, and suburbs are classified as urban areas. Typically, urban areas have high population density and rural areas have low population density (Fig. 1.1).

Different countries have varying definitions of *rural* for statistical and administrative purposes:

In Canada, *Urban area* is defined as having a population of at least 1000 and a density of 400 or more people per square kilometer. All territory outside an urban area is defined as *rural area* (Fig. 1.2).

About 60 million, or one in five Americans, live in rural America. In America, *rural areas* defined by the official Census Bureau classification are sparsely populated, have low housing density, and are far from urban center (Fig. 1.3). Urban areas make up only 3% of the entire land area of the country but are home to more than 80% of the population. Conversely, 97% of the country's landmass is rural but only 19.3% of the population lives there [1].

Brazil does not have a national parameter to define the rural areas. The rural areas are defined administratively by Brazilian municipalities. Rural areas are any place outside a municipality's urban development.

In France and Germany, the rural areas are known as the localities situated outside the urban areas (Fig. 1.4). About 15% of French population lives in rural areas, spread over 90% of the country. Germany is divided into 402 administrative districts, 295 rural districts, and 117 urban districts.

In Britain, "rural" is defined by the Department for Environment, Food and Rural Affairs (DEFRA), using population data from the latest census. The rural area falls outside of settlements with more than 10,000 resident populations.

China has the world's largest population (1.42 billion), followed by India (1.35 billion). The rural population (% of total population) in India was reported as 65.53% in 2019, according to the World Bank collection of development indicators. The National Sample Survey Organization (NSSO) defines "rural" as follows: "an area with population density of up to 400 per square kilometer. The village should have clear surveyed boundaries, but no municipal board. A minimum of 75% of male working population involved in agriculture and allied activities". Researve Bank of India defines rural areas as those areas with a population of less than 40,000. Indian villages have a population of fewer than 500, while 3976 villages have a population of

Healthcare Strategies and Planning for Social Inclusion and Development. https://doi.org/10.1016/B978-0-323-90447-6.00001-1

Rural housing complex in Australia

Rural housing in South Africa

Rural housing in India

Rural housing in Japan

FIG. 1.1

Showing the rural area of different countries located outside the urban area.

FIG. 1.2

Showing landscape of rural area, Canada.

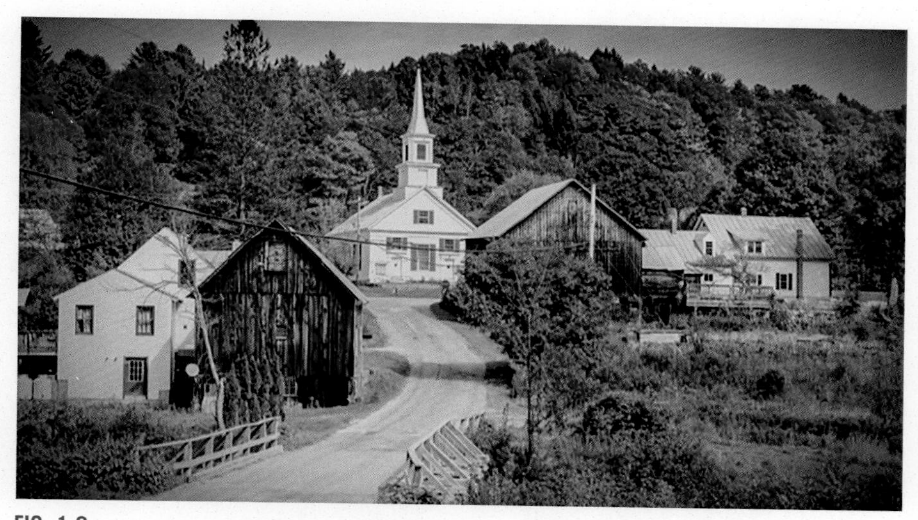

FIG. 1.3

Showing landscape of rural area, America.

FIG. 1.4

Showing landscape of rural area, France.

10,000+. Most of the villages have their own temple, mosque, or church, depending on the local religious following. Rural houses in India are mostly made of nondurable materials taken from the locality (Fig. 1.5).

In Japan, rural areas referred to as "Inak" (the countryside or one's native village). The rural areas refer to all nonurban areas (Fig. 1.6).

FIG. 1.5

Showing landscape of a rural village in India.

FIG. 1.6

Showing rural landscape in Japan.

For policymaking purposes, *urban area* is defined based on population density and the percentage of densely inhabited. Areas that do not meet the population density threshold are rural.

In Pakistan, the rural area is defined as an area that does not come within an urban boundary. According to the 2017 census, about 64% of Pakistanis live in rural areas.

1.2 Access to rural women

For sustainable development in rural areas, the women workforce is the key player for achieving the transformational economic, environmental, and social inclusion. But limited access to credit, healthcare, and education are among the many challenges they face which further worsen due to sudden climate changes, economical and food crises.

So, the only solution for such crises is to empower rural women and involve them in administrative and planning, as decision-maker in state policy finalization. It is necessary to frame well-planned policy at state and national levels to prevent violence against women, prosecuting the culprits before legal authority for justice followed by action. In addition, opportunity should be provided for women's economic independence and should facilitate them with all types of disease preventive measures while involving them in handling health risk measures during pandemic situation like COVID-19.

1.2.1 Rural women as community resource person

Rural women constitute one-fourth of the world's population [2]. They are closely associated with various agricultural practices as labor force (Fig. 1.7), and perform most of the unpaid care work in rural areas [3].

They are the basic workforce for developing community economy which is ultimately responsible for national economy. Mostly, rural women manage the agricultural activities as paid laborers or cultivators doing labor on their own land. The types of agricultural activities taken up by women include: sowing, crop plantation, weeding, irrigation, fertilizer application, and other works related to crop management and development (Fig. 1.8).

Rural livelihood is entirely based on the support and sacrifice of rural women. Rural women play a key role in structure and functioning of rural community (Fig. 1.9).

Total household work including food and nutrition security, generating income, and overall well-being of family and community are well managed by rural women. But rural women and girls regularly confront the problem of social exclusion, and restriction from fully enjoying their human rights, and health disparity.

1.2.2 Rural women as health workforce

To provide access to quality healthcare in rural areas, it is necessary to form well-educated and trained women health workforce with basic knowledge in primary healthcare.

FIG. 1.7

Low wage paid rural women working as daily wages in house construction, India.

FIG. 1.8

Rural women in India involved in crop plantation.

FIG. 1.9

Low wage paid rural women working nearby a rural village, India.

They should be culturally competent with professional license or diploma certificate. Strategies for increasing the efficacy of professional workforce in rural areas include:

- Engaging interprofessional teams to coordinate appropriate care for patients,
- Developing confidence that all professionals are sincerely involved in training and allowed scope for further practice,
- Removing state and federal barriers to professional practice, where appropriate, and
- Allowing to change policy for expressing the existing scope of practice if it is ensured that healthcare workers can provide comparable or better care.

Still, in many developing countries, the rural villages have inadequate health workforce, especially women health providers. In Bangladesh, major deliveries take place at home (62%), and more than 56% of deliveries are assisted by traditional birth attendants (TBAs). In Bangladesh, 36% women do not receive any antenatal care from medically trained persons, and the situation is much worse in rural areas [4].

Almost 94% of maternal deaths occur in low- and middle-income countries (LMICs), including India [5]. India has reported 45,000 maternal deaths in 2015. It is mainly due to lack of female health workforce, socioeconomic and policy-level factors that influence institutional or home delivery without skilled care [6,7]. The only solution for such problem is to train educated rural women, as healthcare provider to meet maternal health, and refer the case to nearby primary health center, in case of any emergency.

Globally, irrespective of the status of a country, rural women have been encountering less access to healthcare than urban women. This problem is more common in the rural areas of developing countries [8], or in least developed countries [9].

Many rural areas have limited number of healthcare providers, especially women health providers. This is mainly due to multiple barriers like geographic location, poor infrastructure for communication, unfavorable weather condition, and inadequate financial resources and specialty healthcare services [10]. In addition, rural localities are restricted access to online information technology compared to urban residents [11], particularly online access to healthcare providers [12].

In addition, women are deprived of access to other productive resources and services. Rural women should access to legislation and policies, decentralize administrative and institutional capacities, and public awareness campaigns need to assert, protect, and enhance rural women's right to health services. Rural women are deprived of accessing financial transition in bank as provided to male; social exclusion in cultural and other social activities; developing any enterprises; and lack of ownership over asserts that can be used as collateral to leverage loans.

1.2.3 Rural women resource to manage energy and drinking water

For sustainable health management and promotion, it is primarily important for a country to have well-managed water resource system. So, the United Nations Environment Programme (UNEP) is working to develop a coherent approach for measuring water-related issues through several multilateral environmental agreement and research bodies. So, Sustainable Development Goals go beyond drinking water, sanitation, and hygiene to also address the quality and sustainability of water resources, which are critical to the survival of people and the planet. The SDGs 2030 goal is to centralize water resources and improve drinking water quality management, sanitation, and hygiene, including health, education, and poverty reduction.

Due to lack of nonavailability of infrastructure and facilities, still, rural women spend an enormous time and energy to fetch water (Fig. 1.10A) and fuel wood for domestic (Fig. 1.10B) and agricultural use, preparing food, and other household work purposes.

FIG. 1.10

(A) Showing how rural women from African village and Sindh (Pakistan) fetch water from a distance for drinking and household work. (B) Showing how rural women from Africa and India fetching fuel wood from nearby forest.

1.2.4 Women for livestock management

Livestock is one of the important parts of global food production system, mainly of rural origin. It is the key sector for the production of commodity for human well-being. It generates food, income, employment, and health factors to mankind. Socioeconomic development of a country cannot be ignored without the multifaceted rural livestock resources. Besides the development of a country, livestock resources are key for the generation of rural employment and supporting rural community from all aspects of livelihood management. For an instant, in Kerala (a highly literate state of India), about 50% of the rural households own livestock and majority of livestock owing households are small, marginal, and landless households. Domestic animals like sheep, goats, pigs, and poultry are largely kept by the land scarce poor householders for commercial purposes due to low initial investment and operational cost.

In rural sector, livestock is the primary effort for livelihood to meet food and shelter, and need of finance for healthcare management and promotion. So, in country like India, it is a noble practice to give animals as a part of women's dowry. Mostly, rural women could manage extra income from the sale of milk and animals. Most of rural women are busy in cattle management activities such as: cleaning of animals and sheds, watering of cattle, fodder collection, preparing dung cakes, and collection of farm yard manure (Fig. 1.11).

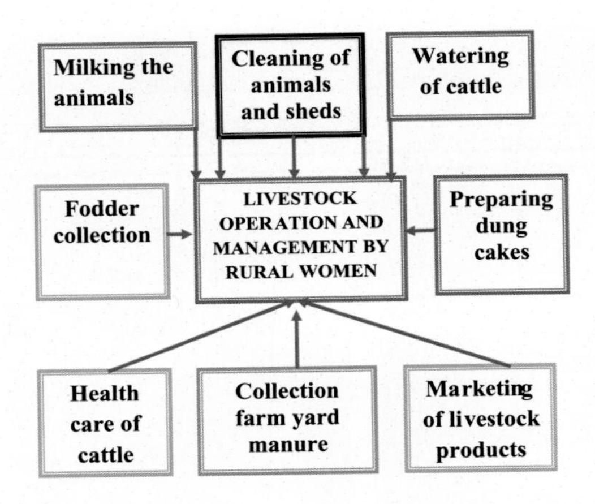

FIG. 1.11

Livestock management and operation by rural women.

In developed country like United States, Food and Agriculture Organization (FAO) enables governments to close gender gaps in agricultural productivity, food and nutrition security, rural livelihoods, and natural resource management, except grazing all other livestock management activities are predominantly performed by women. Men, however, share the responsibility of taking care of sick animals. It is evident that women are playing a dominant role in livestock production and management activities.

Including livestock, forestry, and fisheries and postharvest value addition. The International Fund for Agricultural Development (IFAD) works with governments and other implementation partners to empower poor rural women and development of primary health infrastructure to support economic empowerment of rural women farmers. The United Nations Conference on Trade and Development (UNCTAD) supports developing countries in the integration of gender considerations in their policy formulation and implementation. The United Nations Development Programme (UNDP) helps and supports the government to give guarantee in framing policy, and full legal support as human right.

1.2.5 Women for food security and poverty eradication

Women play critical role in rural structure and functional development. Rural women efficiently and effectively manage rural livelihood by supporting their households and community in achieving food and nutrition security, generation income, and improving rural livelihoods and overall well-being. They contribute to agriculture and rural enterprises and fuel local and global economies. Transformational economic, environment structure and function stability, and preservation of social integrity in rural areas of developing countries are mainly depend on rural women. So, Sustainable Development Goals, especially SDG3 sub-target is meanly benefit for rural women health promotion, by the end of 2030. But, due to limited access, it has been a matter of challenge to empower rural women to look after rural health promotion. Empowering rural women is essential, not only for the well-being of individuals, families, and communities but also for overall economic productivity, given women's large presence in the agricultural workforce worldwide.

At international level, UN Women campaigns and supports the leadership and participation of rural women in shaping laws, policies, and programs on all issues that affect their lives, including improved food and nutrition security and better rural livelihoods.

If opportunity is given, rural women workforce can be helpful in promoting rural socioeconomic status in many ways. Globally, with special reference to developing countries, about 43% of women agriculture laborers belong to rural communities. This could promote the agriculture output of 34 developing countries to an estimated average of 4%. This would be helpful in reducing the number of undernourished people in those countries by as much as 17%, and translating to 150 million fewer hungry people.

Due to lagging of economic growth, hunger is in increasing order in many developing and least developed countries. The annual UN report also found that income inequality is raising in many countries where hunger is on the rise.

1.2.6 Solution for women's disparities

At global level, UN Women supports the leadership and active participation of rural women in making policy decisions at state or central level, framing or bringing amendment in existing laws related to women's right and welfare, and improving food and nutritional activities. In addition, UN Women takes care of providing training for new livelihoods and adapts technology to their needs. In India, UN Women works closely with the Government of India and civil society to set national standards for achieving gender equality. UN Women works to strengthen women's economic empowerment through support to women farmers and manual scavengers.

UN Women helps women from the Tonga ethnic group in Zimbabwe by providing new equipment and training to promote fishing industry along Zambezi River. At present, Zimbabwe women jointly manage the fish marketing in neighbor towns and cities, and participate in a revolving small loan.

Earlier, in China, women farmers have limited access to irrigation technology. UN Women helps women farmers of Ningxia Hui Autonomous Region how to acquire and maintain advanced irrigation system, according to climatic condition. Due to agricultural production problem, many men left village to nearer cities or towns for daily livelihood. Under such condition, women started learning the irrigation technology with the help of UN Women and could able to manage agricultural productivity.

The Food and Agriculture Organization, the International Fund for Agricultural Development, and World Food Programme extended their help and financial support to empowered rural women to claim their rights to land, leadership, opportunities, and choices, and to participate in shaping laws, policies, and programs. Evidence shows that this spurs productivity gains, enhanced growth, and improved development prospects for current and future generations. The initiatives also engage with government to develop and implement law and policies that promote equal rights, opportunities, and participations so that rural women can benefit from trade and finance, market their goods, and make a strong contribution to inclusive economic growth.

1.3 Rural women health disparity

Since the recent past, worldwide, much discussion has been going on about the inequities between rich and poor and between majority and minority groups. In addition, the world community is keen on how to resolve this issue and bring down the problem of health risks persisting among racial and ethnic groups. Attention also need to be paid to the disparities between women who live in rural areas and those who live in urban part of the state. Rural women are managing their livelihood under the deficit economic budget without any health insurance than their urban counterparts. So, it is high time to find out the root cause of health disparities between women residing in rural and urban parts of the state and explore strategies to mitigate them that include adequate healthcare workforce, improved primary healthcare infrastructure and provision of telehealth services, and expanding health insurance option.

1.3.1 Women health disparity in COVID-19

The COVID-19 pandemic widely affects healthcare more than any other sector. Women are 70% of healthcare workers globally, and they are 73% of the healthcare workers who have been infected with SARS-CoV-2. There are concerns about the impact of COVID-19 on pregnancy, with some hypothesizing that it may cause placental issues due to its effect on blood clotting.

Mostly, the Personal Protective Equipment (PPE) is not designed accordingly for women's body. As a result, women health workers deal with poorly fitting face masks and the need to remove their PPE more often during long shift, for menstrual hygiene.

Women are serving on the frontlines against COVID-19, and resulted in high risk of health problem. Women are over burden with additional health care, continue to do the majority of unpaid care work in households, face high risk of economic insecurity, and face increased risk of violence, exploitation, abuse, or harassment during times of crisis and quarantine raised due to COVID-19.

From medical perspective, at the beginning stage of COVID-19, men fatality was 60–80% higher than women. But, at the later stage of pandemic, the impact of the pandemic on women is becoming increasingly severe. This is mainly due to the service of women that is at the forefront of the battle against the pandemic as they make up almost 70% of the healthcare workforce, exposing them to greater risk of infection, while they are underrepresented in leadership and decision-making process in the healthcare sector. Women health workforce constitutes two-thirds of the health workforce worldwide. Globally, they are underrepresented among physicians, dentists, and pharmacists but they came up to around 85% of nurses and midwives in the 104 countries for which data are available [13].

Women also dominate over the majority of the long-term care (LTC) workforce around the world, valuing 90% of the total global workforce, both men and women (Fig. 1.12). In spite of majority of workforce belonging to female community, women still make up only a minority of senior or leadership positions in health [14].

Travel restriction and quarantines, closer of day care center for children, imposition of additional burden by elderly member at home are expected to increase unnecessary worry for house wife, even though she remains confined with her partner and working together.

1.3.2 Gender equality in COVID-19

Recently released report from UN Women (the United Nations Entity for Gender Equality and the Empowerment of Women) shows that the pandemic will result in 96 million people into extreme poverty by the end of 2021, out of which 47 million of whom are women and girls. This will bring the total number of women and girls living on USD 1.9 or less to 435 million [15].

COVID-19 pandemic resulted in increase in gender equality and poverty problems at global level. In the end of 2021, more women will be pushed into extreme poverty than men. UN Women through its global gender data programme Women Count, in partnership with UNFP, the 28-country study, revealed that the problem

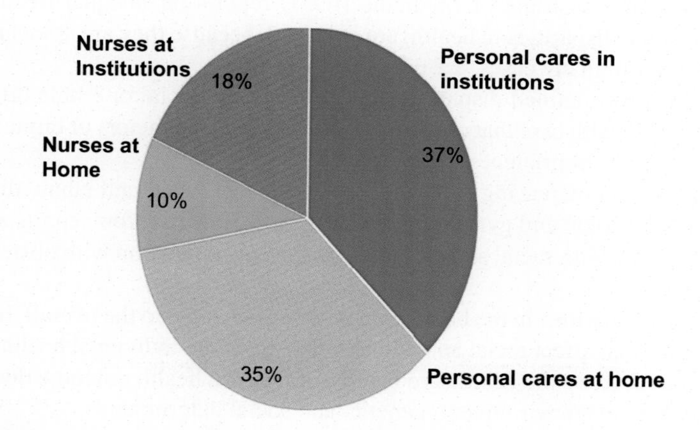

FIG. 1.12

Over 70% of LTC workers are personal carers across OECD (Organisation for Economic Co-operation and Development) countries.

Source: EU-Labour Force Survey and OECD Health Statistics 2018 (data refer to 2016 or nearest year).

on gender equality will be worse during post-COVID-19. The study reports that more than 60% of women and men in Ethiopia, Kenya, Malawi, Mozambique, and South Africa have worst experience of abrupt loss of personal incomes due to the COVID-19 pandemic.

The sudden reduction in income has badly affected food security in Ethiopia, Kenya, and South Africa. About one in five people is victimized for a day or more. Education system also collapsed during the lockdown period due to COVID-19 pandemic. About 25 million students are affected in Ethiopia followed by South Africa with 15 million, and Kenya and United Republic of Tanzania with 14 million learners affected each.

From health point of view, with little exceptional country, women are more victimized as compared to male population.

1.3.3 Racial and ethnic disparities for women

Racial and ethnic disparities in healthcare are commonly noticed in most of the developing countries. Even developed country like USA has great racial and ethnic disparities in white and non-white population, especially with reference to obstetrics and gynecology. People from racial and ethnic face many challenges when it comes to medical care, and they often have great need for medical care owing to higher levels of morbidity and comorbidity (the simultaneous presence of two or more diseases). Many racial/ethnic minority populations have lower levels of access to medical care in United States than do whites.

In spite of remarkable advanced research and development of medical technology, still the ethnic minorities tend to receive lower quality of care than nonminorities, and resulted in greater morbidity and mortality from various chronic diseases than nonminorities. The Institute of Medicine (IOM) reports on unequal treatment as: "racial and ethnic disparities in healthcare exist and, because they are associated with worse outcomes in many cases, are unacceptable" [16].

The IOM report defined disparities in healthcare as "racial or ethnic differences in the quality of health care that are not due to access related factors or clinical needs, preferences, and appropriateness of intervention."

There is also an increasing recognition of persistent racial and ethnic disparities prevalent in obstetrical and gynecological outcomes. However, few reports are available in connection with racial and ethnic disparities in connection with obstetrics and gynecology.

In general, disparities in the healthcare system contribute to the overall disparities in health status that affect racial and ethnic minorities. The individual health status is not only influenced by the healthcare but also a resulted health outcome due to various other factors like environment, genetic, and social determinants.

For instance, about 38% of US females are belonging to ethnic minority group, or both. As of 2013, about one-half of US birth was to women of color, and it has been predicted that non-white population will be major as compared to white color individuals by the end of 2050. Infant mortality rates among minority groups lag behind those of whites. A decade back, among African Americans, the infant mortality rate was 2.4 times greater than that of the white population [17].

Office of Minority Health (OMH) at the US Department of Health and Human Services (HHS) reports [18–20]:

- Non-Hispanic blacks/African Americans have 2.3 times the infant mortality rate as non-Hispanic whites.
- Non-Hispanic black/African American infants are four times as likely to die from complications related to low birth weight as compared to non-Hispanic white infants.
- Non-Hispanic black/African American infants had twice the sudden infant death syndrome mortality rate as non-Hispanic white, in 2018.
- In 2018, non-Hispanic black/African mothers were twice as likely to receive late or no prenatal care as compared to non-Hispanic white mothers.

It is high time to understand unrecognized racial and ethnic disparities in women's health, and immediately take action to identify and eradicate these disparities.

The National Institutes of Health says, "health disparity population can be identified on the basis of disparity in the rate of disease incidence, occurrence of morbidity or mortality in a specific population as compared with general population" [21].

In general, health disparities are commonly observed in many groups, but it is noticed to be more among women who are members of racial and ethnic minority groups. The cases of obstetrician-gynecologists are most prominent as compared with other health issues [22].

The conceptual development of race and ethnicity is mainly due to social rather than biological origin which can focus on about how environmental, cultural, behavioral, and medical can affect patients. Sometimes, the frequencies on genetic variations bring the issues in different racial or ethnic groups. For example, genetic polymorphisms associated with increased susceptibility to disease also may vary in frequency in different racial and ethnic groups. Primarily, race and ethnicity are social constructs, the effect of common ancestral lineage on the segregation and frequency of genetic variations in combination with the influence of cultural factors which are ultimately responsible for health disparities.

Although awareness on racial/ethnic disparities has been increased among policymakers, still, there is little consensus on what can or what should be done to reduce these disparities. The following are few measures to reduce racial and ethnic disparities:

- By developing awareness among healthcare providers and administrators about the racial and ethnic disparities in a community.
- To make realize the health workforce to have nonbiased attitude toward racial groups while tendering health services.
- To strongly encourage the adoption of federal standards for collection of race and ethnicity information in clinical and administrative data to better identify disparities.

- To encourage research to identify structural and cultural barriers responsible for racial disparity while serving the people for health promotion, and also to ensure the effectiveness of interventions to address such barriers.
- Educating patients in a culturally sensitive manner about steps they can take to prevent disease conditions that are prevalent in their racial and ethnic groups.
- Encouraging in assisting recruitment of obstetrician-gynecologists and other healthcare providers from racial and ethnic minorities into academic and community healthcare fields.

1.4 Women health and UN women act

Everyone has the right to a standard of living adequate for the health and well-being of himself and his or her family, including food, clothing, housing, medical care, and necessary social services, and the right to security in the event of unemployment, sickness, disability, widowhood, old age, and other related issues on health promotion.

Although women and men have different healthcare needs, they are entitled for an equal right to live healthily. Due to financial constrain and social exclusion, many women and girls encounter gender discrimination which ultimately resulted in morbidity and mortality. In 2017, the global martial mortality rate (MMR) was enormously high, numbering about 95,000 women death following pregnancy and childbirth [23].

This incident was followed by Sub-Saharan Africa and southern Asia with death tools about 86% (254,000) of the estimated global maternal death. Sub-Saharan Africa alone accounted for about two-thirds (196,000) of maternal deaths, while southern Asia accounted for nearly one-fifth (58,000).

The MMR rate was surprisingly reduced between 2000 and 2017 to nearly 60% (from an MMR of 384 down to 157). In spite of its very high MMR in 2017, Sub-Saharan Africa as a subregion also achieved a remarkable decline in MMR of about 40% since 2000. During the same period, significant decline in MMR in subregions: Central Asia, East Asia, Europe, and North America was observed. Overall, the maternal mortality ratio (MMR) was less in less-developed countries, declined by just under 50%.

The main cause of high maternal mortality deaths in some region is mainly due to inequalities in access to quality health service and presence of significance gap between rich and poor. The MMR in low-income countries in 2017 is 462 per 100,000 live births versus 11 per 100,000 live births in high-income countries. In 2017, Fragile States Index disclosed about significantly high MMR in 15 countries (South Sudan, Somalia, Central Africa, Yemen, Syria, Sudan, the Democratic Republic of the Congo, Chad, Afghanistan, Syria, Sudan, Zimbabwe, Nigeria, Ethiopia, and Haiti). The adolescent girls under 15 years are prone to complication in pregnancy and childbirth as compared to women aged 20–24 [24,25]. In less developed countries, the number of pregnancy is more as compared to developed countries. Most of the death cases occur due to complicacy developed during pregnancy, although most of the cases can be preventable or treatable. The major complicacy symptoms in pregnancy that occurs in about 75% of all maternal deaths are [26]:

- Pre and post bleeding during childbirth
- Post infection to mother after childbirth
- High blood pressure during pregnancy (preeclampsia and eclampsia)
- Complication from delivery
- Unsafe abortion

Most maternal deaths are preventable, as most of the healthcare solutions to prevent or manage complications are well known. It is essential to have well-managed healthcare system, especially for women for proper and timely treatment before and after the childbirth. Both the maternal health and newborn baby's healthcare are equally important. So, all the delivery cases should be attended by skilled health professionals, as timely management and treatment can make difference between life and death for mother as well as for baby.

Still, the rural inhabitants are depriving of primary healthcare than their urban counterparts. This is real fact for the regions with low number of skilled health workers, such as Sub-Saharan Africa and South Asia.

The health of women and girls is of particular concern because in many societies, they are suffering from discrimination rooted in social-cultural factors. For example, women and girls face increased vulnerability to HIV/AIDS.

Some of the important social-cultural factors mainly act as barriers for women and girls benefit from quality health services include:

- Men and women are unequally differentiated in possessing power.
- Many of the social exclusions keep women and girls away from basic educational facilities and deprive them for employment opportunity.
- Persistence of physical, sexual, and emotional violence in rural communities.

Poverty is an important barrier to positive health outcomes for both male and female, especially for women it resulted in high burden of unsafe cooking practice (directly use of firewood) and feeding undernutrition leading to the problem of malnutrition.

The people with higher income residing in urban areas used to avail more than 90% of all birth benefits from the presence of well-trained midwife, doctor, and nurse. However, fewer than half of all births in several low-income and lower-middle-income countries are assisted by such skilled health providers [27]. Factors like poverty, distance to facilities, lack of telehealth facility, inadequate availability of health workforce, and cultural belief and practices are acting as hurdles for discharging healthcare services, timely.

In order to overcome various issues related to health risks, the United Nations Entity for Gender Equality and the Empowerment of Women (UN Women Act) was established for working as a United Nations entity for the empowerment of women. The UN Women became operational in January 2011.

The main target of UN Women is to work with governments to improve health services for women and girls, including survivors of violence. In addition, UN Women actively concerns to stop discriminatory laws and practices impeding women's access to sexual and reproductive healthcare services. UN Women's main thematic areas of work include:

- Leadership and political participation
- Economic empowerment
- Ending violence against women
- Humanitarian action
- Peace and security
- Governance and national planning
- Gender equality
- The 2030 Agenda for Sustainable Development

UN Women with the global partnership with Sustainable Development Goals program has focused on the following social development program at global level:

- By 2030, reduce the global maternal mortality ratio to less than 70% per 100,000 live births.
- By 2030, ensure universal access to sexual and reproductive healthcare services, including family planning, information and education, and the integration of reproductive health into national strategies and programs.
- By 2030, end preventable deaths of newborn and children under 5 years of age, with all countries aiming to reduce neonatal mortality to at least as low as 12 per 1000 live births and under-5 mortality to at least as low as 25 per 1000 live births.
- To extinguish life torture epidemics like AIDS, tuberculosis, malaria, and neglected tropical diseases and to combat hepatitis, waterborne diseases, and other communicable diseases like COVID-19, by the end of 2030.
- To minimize premature mortality through preventable and curable methods, from noncommunicable diseases by one-third rate of the present victimization on premature mortality by the end of 2030.
- To treat and prevent abuses on narcotic drug and use of harmful alcohol.
- To reduce the number of causalities caused due to the unsafe use of hazardous chemicals by 2030.
- To enhance campaign against use of tobacco control as sponsored by World Health Organization.
- To encourage and support the research and development of affordable vaccine and medicines for the communicable and noncommunicable diseases, as per Doha Declaration on the TRIPS Agreement and Public Health. This declaration was adapted by the World Trade Organization (WTO) Ministerial Conference of 2001 in Doha on 14 November, 2001. It reaffirmed flexibility of TRIPS member states in circumventing patent rights for better access to essential medicines.

To provide the facilities of early warning to all countries, especially to developing countries on health risk reduction and management of national and global health risks.

1.5 Rural Women's health

WHO has defined health as "a state of complete physical, mental, and social well-being and not merely the absence of disease or infirmity" [28]. Health can influence an individual's ability to reach his or her full potential in society.

WHO has defined health disparity as "the differences in health care received by different groups of people that are not only unnecessary and avoidable but unfair and unjust" [29].

Health disparities are noticed as universal phenomenon in the all type of income groups. There are wide disparities in health status of different social groups. The lower an individual's socioeconomic position, the higher their risk of poor health status of different population groups. Social factors like education, employment status, income level, gender, and ethnicities have significant influence on how healthy a person is. Health disparities exist irrespective of a country as low-, middle-, or high income. There are wide disparities in health status of different social groups. The lower an individual's socioeconomic position, the higher their risk of poor health status. Social factors like education, employment status, income level, gender, and ethnicities have significant influence on how healthy a person is. Both males and females are victimization of health disparities. But girls and women experience a majority of health disparities. This is mainly due to parochial, social, and cultural practices in most of the rural areas of developing or underdeveloped countries. This short of practice results in mistreatment and abuse of women that are ultimately responsible to make women more prone to illness and early death [30].

1.5.1 Urban women vs. rural women

Rural women face poorer health outcomes and confess inadequate healthcare facilities as compared to urban women. Rural areas have extremely poor healthcare providers, especially women's health providers (Fig. 1.13).

Even in developed country like America, rural American women face heterogeneous health disparity depending on the geographical location and socioeconomic condition. The present population (June 2021) in America is 333.6 million, out of which male are 49.4 and female are 50.6 (Source: United Nations Department of Economic and Social Affairs: Population Division).

India is the second biggest populated country. The total population in India (June 2021) is 139.8 billion out of which 51% are male and 48.43% female (Source: United Nations Department of Economic and Social Affairs: Population Division). Women in India face heavy gender biases and are subsequently more likely to experience disadvantages in their lives related to healthcare.

Women are responsible for sustainable development and quality of life in the family. But women face more health problems than men, and consume more prescribed drugs as compared to men. Women feel desperate when visiting to physicians, as they keep them waiting for long time for health checkup and prescribing medicine. This is the reason why rural women prefer local low-qualified health profession for

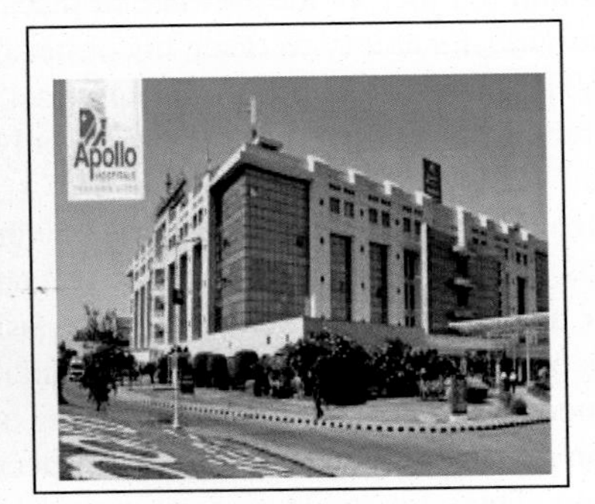

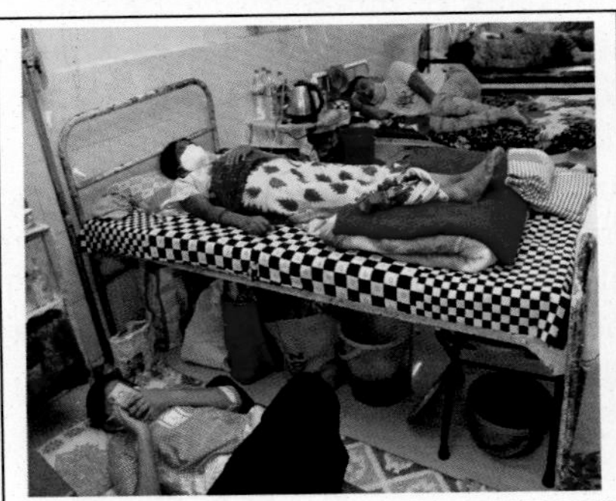

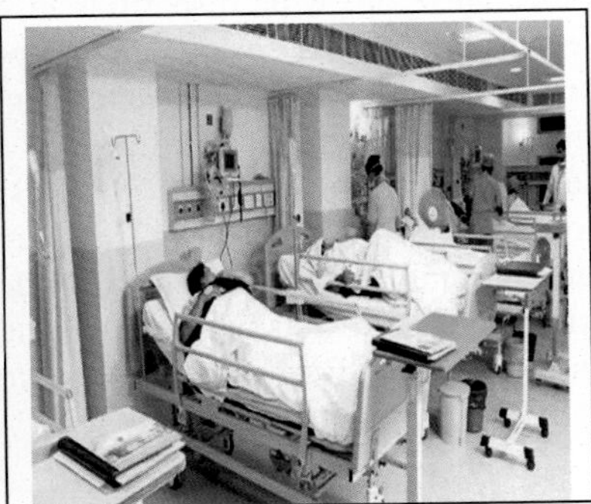

FIG. 1.13

Poor infrastructure for health services in rural areas of developing country, India.

treatment, and often suffer seriously due to wrong diagnosis. This is mainly due to the living style (Fig. 1.14), lack of poor access to infrastructure facilities, low economic budgetary provision, and poor attention from government that significant health disparities exist between rural and urban women.

Global health-related issues (life expectancy, sanitation, health coverage, etc.) are mainly based on multiple factors including geography, socioeconomic status, and race/ethnicity age. These factors are directly or indirectly responsible for significant health disparities that exist between rural and urban women. Various definitions of "rural" are used to study and report population data and to determine eligibility and reimbursement level for federal and state programs.

Due to the limited availability of data on rural women health, it has been difficult to find out challenges on health disparities with special reference to rural locality. Even in developed country like the US, rural women experience high rates of disparity than the urban counterparts. It has been noticed that about 20% of the US population residing in rural areas [31,32] confront with the problems like higher rate

FIG. 1.14

Disparities in living style of urban and village women.

of mortality [33], premature morbidity, and mortality life-threatening diseases like cancer, heart disease, and childhood obesity, and abnormal behavior [34–37] due to lack of preventive healthcare services [38,39]. However, the European Institute for Gender Equality has developed a Gender Equality Index which can be helpful in measuring to compare among the member states (Fig. 1.15).

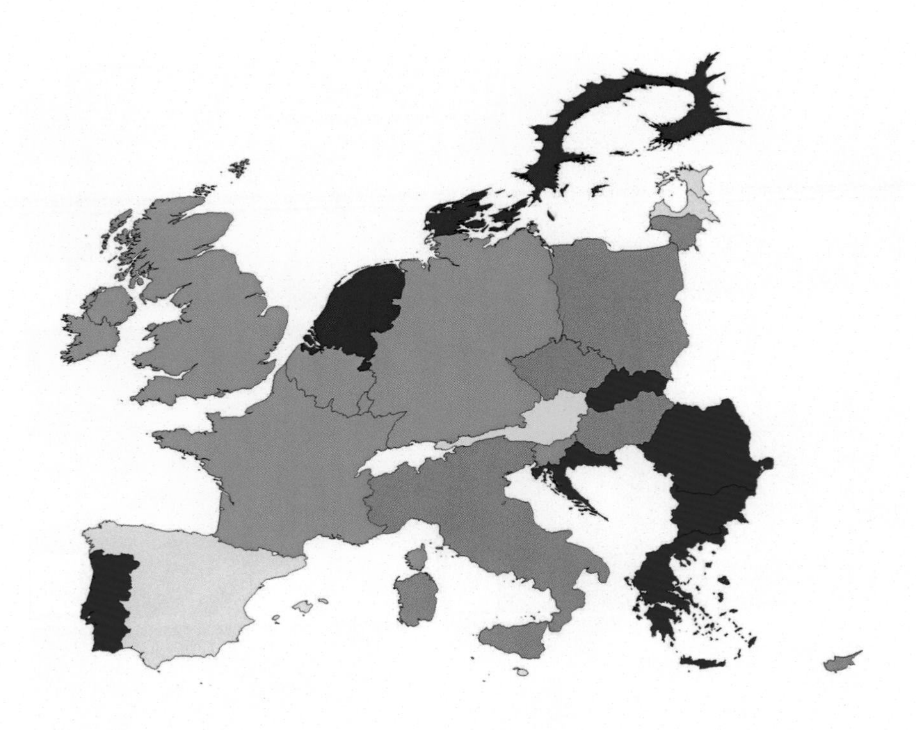

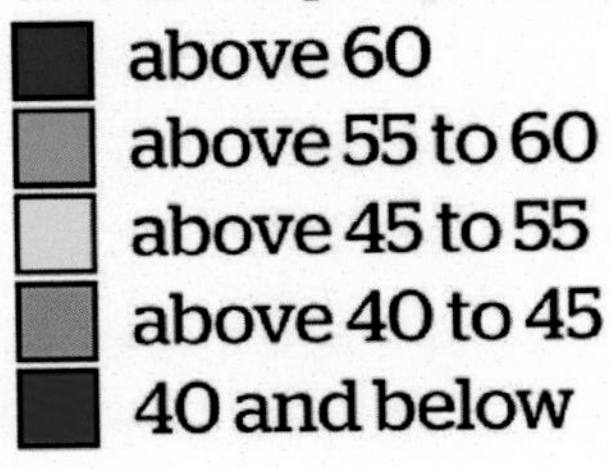

FIG. 1.15

Gender equality index score by European Institute for Gender Equality (1 = total inequality, 100 = full equality).

In addition, in the US, injury by motor vehicle-related deaths and cerebrovascular incident of cervical cancer [40,41,41a] are noticed, but ignored due to nonavailability of emergency healthcare facilities.

In developing country like India, working women in urban parts of the country are involved in household chores more than their counterparts in rural areas.

According to 2011 National Sample Survey (NSS), rural women make up 81.29 of the female workforce in India. Most of these women are agricultural laborers who work on someone else's land in return for wages. About 56% of working rural women is illiterate, whereas the number of illiterate urban working women is 28% lower. But, under the present situation and political stability in India, labor women workforce participation rate has been declined from 42% in 1993–94 to 31% in 2011–12. The World Bank research group claims that the drop in the participation rate among rural girls and women aged 15–24 is the recent expansion of secondary education and rapidly changing social norms leading to "more working young age females opting to continue their education rather than join the labor force early."

1.6 Women's health disparity in different countries

Irrespective of the status of a country, gender-based health disparities have been a common problem throughout the world. Women in developing country and least developed country are worst victimized with the problem of health disparity. Due to lack of basic healthcare services, women face life-debilitating and life-threatening health issues. Following are few countries showing the problem of health disparity in women, mostly in rural areas:

1.6.1 Health disparity in European women

Europe has taken first status on gender equality in the world but with slow progress. Some states of Europe have not yet followed gender equality. According to existing gender equality indexes, no Member of States has achieved full equality.

Last five decades have witnessed progress in women's equality in most of the places of Europe (Fig. 1.16). The European countries have implemented initiatives to attain a more gender-balance workforce with the introduction of family-friendly policies. In academia, however, fewer women reach top leadership positions than those in the political arena. In December 1995 at Madrid, the European Council Summit supported the resolution of the Beijing + 15 on gender equality and necessary initiation was made to adapt the goals of the Beijing platform for action.

The Gender Equality Forum is a major global inflection point for gender equality. These landmark victories process ahead with government effort, and corporation voluntary bodies to provide opportunity for women and girls worldwide.

European government now took priority for gender equality as part of Beijing +25 review. On the basis of SDG3 targets, the members of state planned to implement gender equality by the end of 2030 for sustainable development (Fig. 1.17).

In September 2020, European Member of States reviewed entire protocol on gender equality and its impacts on both women's and men's health as part of monitoring progress under the Strategy on Women's Health and Well-being in the WHO European Region (2016) and the Strategy on the Health and Well-being of Men in the WHO European Region (2018).

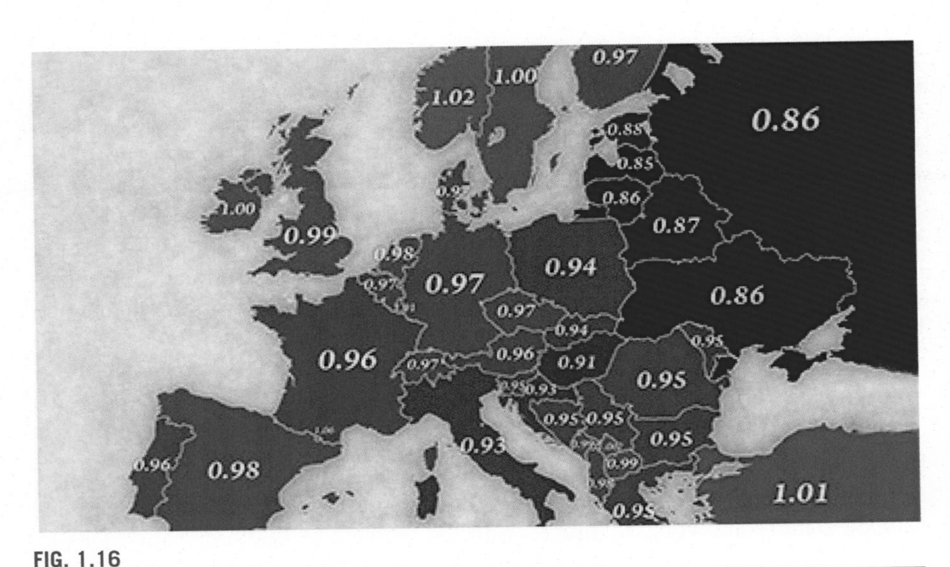

FIG. 1.16

Gender inequality in the European Union.

Source: www.viewsofftheworld.net.

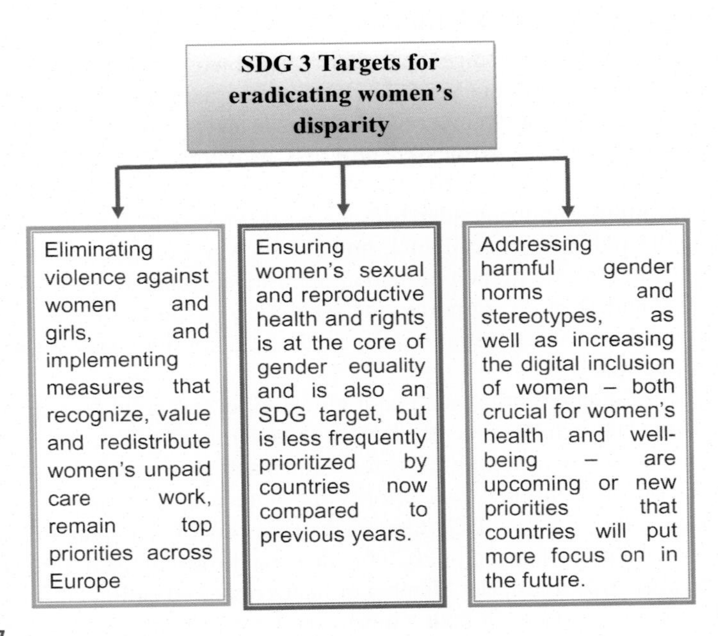

FIG. 1.17

Planning of SDG3-related target implementation in the States of European Country.

Fortunately, in March 2020, about 19% of health ministers in European Member of States are women. But this is down from 34% in August 2019. Yet, about 90% of the female are long-term health and social care workforce in countries of the Organization for Economic Co-operation and Development. Women are responsible for majority of the health workforce in the European region, yet gender inequalities within health system persist. Female health workers still face significance barriers in terms of achieving leadership position and income equality and overcoming stereotypes about the healthcare roles that women generally fill. Healthcare services depend mainly on women who contribute as informal care givers, particularly for children and older people.

1.6.2 Health disparity in Asia-Pacific

Access to healthcare was noticed to be increasing, since 2018, but women in low-income households in rural area still have difficulty in accessing care, due to distance and financial reasons. Health at a glance: Asia/Pacific 2018 says that in Cambodia, Nepal, Philippines, and Solomon Islands, more than three women in four with lowest household income are facing difficulties in accessing healthcare due to financial reasons. In Nepal, Pakistan, and Solomon Islands, about two women in three from worst-off households are facing difficulties in accessing healthcare mainly due to distance.

It has been noticed that the expectancy has increased by almost 6 years, since 2000 to reach 70 years across lower-middle and low-income Asia-Pacific countries, but maternal mortality is still twice the Sustainable Development Goal target in these countries. Many countries in this region face a double burden of disease, as they still struggle to reduce maternal and child deaths at a time when the prevalence of chronic conditions and unhealthy lifestyle is growing. More than one-third of adults are overweight in Asia-Pacific, and one in 10 persons is obese. Among the children, 5% of under age 5 and more than 20% of adolescents are overweight.

It has been observed that between 2000 and 2015, the average mortality rate in lower-middle and low-income Asia-Pacific countries has been reduced to half, as compared to earlier data. But still, it persists about 140 death per 100, 000 live births which is twice the value of SDG target of 70 deaths over 1000,000 live births.

Adolescent girls continue to face considerable disadvantages in relation to sexual and reproductive health and right, including protection from child marriage. Girls have poor access to modern contraception and experienced high rates of intimate partner and sexual violence. The social discrimination and exclusion of women and girls are most commonly noticed in South Asia.

At present women play special driving role in economic growth in Asia-Pacific. Rapid industrialization and socioeconomic changes have seen a significant increase in women in the workplace over the past three decades, leading to structural transformation in the labor force and real poverty reduction. Despite this progress, women continue to hold merely one in five position of leadership in the Asia-Pacific workforce, and have been facing continuous obstacles to career progression. According to World Bank, women make up 39% of the labor force, lagging those of men, and in Asia-Pacific around 60% versus 80% for men, women are knowingly ignored for senior position in any governing body of an organization.

The main barrier to women's leadership in Asia-Pacific is the gender inequality and social exclusion in appointing as workforce in different organizations. A female commitment for social responsibility contributes toward limiting women's opportunities. The majority of women engage in approximately 4h daily of unpaid labor. That means both less time available and less income for women, leaving them more vulnerable than men. Therefore, access to the workforce is not a solution in itself to gender inequality.

SDGs target is to pay attention to gender equality as a global human right and health and development priority. The current focus on girl's sexual and reproductive health and elimination of harmful practices is well justified, and data from Asia-Pacific region show that much remains to be achieved. Gender inequality remains as one of the most pervasive challenges in global health development.

1.6.3 Health disparity in Africa

About 2.30 million are in Sub-Saharan Africa (SSA) having the age group between 15 and 49 years. In 2015, the maternal deaths were about 546 per 100,000 live births. According to the data available from the survey of Millennium Development Goals (MDGs), half of women in SSA do not have access to the essential healthcare during pregnancy and childbirths and contraceptive use remains low and insufficient with only 28% prevalence in 2015 among women.

Mainly, women's health is the basic of social and economic development in the African region. Women population in Africa represents slightly over 50% of the continent's human resource. Women's health has huge implication for the region's development.

Due to inadequate service opportunity, huge gap exists between individual belonging to different socioeconomic groups. In this connection, World Bank reports that the concept of equality of opportunity requires that individuals' opportunity are independent of their life circumstances into, such as religion or wealth of one's parent, and over which she has no influence.

It has been observed that maternal and reproductive health opportunities are unequally distributed among population groups with different wealth characteristics, areas of residence of educational level.

1.6.4 Health disparity women in China

Due to family reproduction issues rural women have been suffering a lot. The People's Republic of China has major women health problems. For example, the child mortality rate is 10 times higher in the poorest province than in the richest one and the gap in life expectancy between two provinces is as high as 13 years.

In order to overcome such problem, during 1980s, China had well-planned economy targeting to socioeconomic development, mainly based on healthcare coverage. In early 2000, China attempts to give stress on social health insurance as practiced in urban employ [42–45].

1.6.5 Health disparity in India

Next to China, India is the highly populated country where women face wide range of diseases and health problems which ultimately affect the aggregate economy's output. Depressing the gender, class, or ethic disparities that exist in healthcare and improving the health outcomes can contribute to economic gain through the creation of quality human capital and increased levels of savings and investment. Presently, some states of India provide primary healthcare services such as vaccination, develop temporary healthcare unit, and bring timely awareness among rural woodmen to take care of family and children's health (Fig. 1.18).

Education, economic, and political inequality between women and men in India is in deteriorating condition. Gender inequalities, and their social causes, impact India's sex ratio. Gender inequality in India is a multifaceted issue that concerns men and women. Women are in several disadvantages in India compared to men. Although constitution of India grant men and women equal rights, gender dispraise remains. But, in India, still dowry and adultery like disparity factors are in alarming rate, affecting the lives of many today.

1.6.6 Health disparity in the US

About 22.88% of US women aged 18 years and older live in rural America [46,47]. In America, rural communities are heterogeneous with substantial regional difference in ethnic and racial composition [48,49].

FIG. 1.18

Community health worker preparing a vaccine, India.

Mostly, rural women confront poor health status, unintentional injury, motor vehicle-related deaths, cerebrovascular disease deaths, suicide, cigarette smoking, obesity, difficulty with basic actions or limitation of complex activities [50,51], and incidence of cervical cancer [52]. In addition, the death rates from ischemic heart disease in rural women exceed that for all US women. In some regions of the country, women in nonmetropolitan areas have higher rates of heavy alcohol consumption [50].

Rural women of African American, Hispanic, Asian, and white are desperately lacking preventive screening services for breast and cervical cancer.

From health services access point of view, rural women face lot of disparities in reaching a nearby health services center, timely. About 50% of rural women have transport facility to reach health center in 30 min drive to avail the facility of perinatal services [53]. Within a 60-min drive, the proportion increases to 87.6% in rural towns and 78.7% in the most isolated areas [54,55].

1.7 Health inequality in men and women

The inequality is a comparable term mostly referring to the facts like social exclusion, disparity of distribution or opportunity, or lack of evenness. The practice of social exclusion of women in rural areas of developing counties leads to inequality in women's health status [56–58].

From education point of view, only 66% of countries have achieved gender partially in primary education. At the secondary level, the gap widens: 45% of countries have achieved gender parity in lower-secondary education and 25% in upper-secondary education. This is mainly due to barriers to girl's education like poverty; child marriage and gender-based violence vary among countries and communities. Poor families often favor boys when investing in education. Most of the rural locality and periphery of urban locality have lack of safety and sanitization needs of girls.

Mostly, women's morbidity and mortality is expressed on the basis of their own occupation which is supposed to be less than that of men. The extent of social exclusion in women's health is equal to the way inequality is defined and measured [56,57,58]. The male and female health standards are supposed to be similar on the basis of their respective healthcare coverage. The Office for National Statistics (ONS) suggests to study sex differences in health inequality on the basis of social positions and advantages. In addition, condition of employment and other maternal cultural aspects of lifestyle outside the workplace also to be considered (Table 1.1).

1.8 Women's chronic disease and prevention

Noncommunicable diseases (NCDs) are also known as chronic diseases, tend to be of long duration and are the result of a combination of genetic, physiological, environmental, and behavioral factors. The main types of NCD are cardiovascular

Table 1.1 Gender/sex-related inequalities in term of diseases.

Condition	Gender/sex-related inequality (or difference)
Asthma	Prevalence is higher in women
Autistic spectrum disorder (ASD)	Prevalence is higher in men
Autoimmune disorders	Prevalence is generally higher in women; specific examples include rheumatoid arthritis and multiple sclerosis
Breast cancer	One in eight women will develop it in their lifetime
Asthma	Prevalence is higher in women
Prostate cancer	One in 10 men will develop it in their lifetime
Lung cancer	Lifetime risk is higher in men; incidence is increasing in women
Colorectal cancer	Lifetime risk is higher in men
Chronic liver disease	Mortality and morbidity rates are higher in men
Chronic obstructive pulmonary disease	Mortality higher in men; similar incidence in men and women
Coronary heart disease	Incidence is higher in men
Diabetes	Prevalence is slightly higher in men (for both type I and II combined)
Disability	Prevalence of self-reported limiting long-term conditions is similar in men and women
Epilepsy	Prevalence is higher in men
Hepatitis C	Prevalence is higher in men
Injuries	Incidence is higher in men under 65 years, higher in women of 75 years and over
Mental health	Self-reported prevalence of well-being similar for men and women (16 years and over), but higher prevalence of mental ill health among women
Osteoporosis	Higher prevalence in women
Sexually transmitted infections	Overall, higher incidence in men, but chlamydia incidence is higher in women
Stroke	Incidence rate is higher for men; mortality rate is higher for women
Suicide	Suicide rate is higher for men than women

diseases (such as heart attacks and stroke), cancer, chronic respiratory diseases (such as chronic obstructive pulmonary disease and asthmas), and diabetes. NCDs disproportionately affect people in low- and middle- income countries where more than three-quarters of global NCD deaths (31.4 million) occur.

Irrespective of age, location, and gender, all people are affected by NCDs. Mostly, older age groups are more victimized as compared to younger generation. But WHO report says that more than 15 million of all deaths caused due to NCDs occur between the age of 30 and 69 years. People from the lower- and middle-income countries mainly suffer from the premature death. Generally, children, adults, and elderly people are more prone to risk factors (unhealthy diets, physical inactivity, expose to tobacco smoke, or use of alcohol) contribute to NCDs.

Various physical and chemical factors like rapid unplanned urbanization, globalization of unhealthy lifestyles, and population aging are likely responsible for NCDs. It is evidenced that lack of proper diet and physical activity may be responsible for blood pressure, increase in blood glucose, elevated blood lipids, and obesity. These are known as metabolic risk factors that can make an individual prone to CVDs. Following are few important risks factors responsible for NCDs:

- About 7.2 million deaths per year are due to tobacco use.
- Alcohol use attributes about 3.3 million annual deaths.
- Lack of regular physical exercise causes about 1.6 million death annually.
- About 4.1 million deaths have been attributed to excess salt/sodium intake [59].

One of the major issues of the SDGs agenda is to reduce the premature death due to NCDs by one-third by the end of 2030. As poverty is one of the major factors causing NCDs premature deaths, the SDGs are targeted to reduce poverty in low-income countries. To remove poverty in all its forms is one of the greatest challenges. While the number of people living in extreme poverty dropped by more than half between 1990 and 2015, too many are still struggling for the most basic human needs.

It has been commonly noticed that vulnerable and socially disadvantaged people are more prone to sudden death as compared to people of higher social positions. The exorbitant cost for the treatment of NCDs, which often lasts for a longer period is very high, and is difficult for a common individual to bear the cost from own pocket.

The better way to minimize the risks of NCDs is to focus on the risk factors associated with these diseases. Some of the important measures for reducing risk factors for NCDs include:

- By increasing conscious on international cooperation and camping on vulnerability of NCDs and prevention.
- Bringing understanding and strengthening the international organization, voluntary bodies, and stakeholders to proceed for the prevention of risk factors related to NCDs.
- To bring better amendment in controlling risk factors responsible for tobacco use, harmful use of alcohol, habituated with unhealthy diets, and physical inactivity.
- By implementing more effective legal frameworks to have control over risk factors related to NCDs.
- By bringing necessary amendment in the present health system to cover in better way the campaign for universal health coverage.
- Promote high quality research and development to prevent NCDs.
- Monitor trends, determinants, and progress to achieve global, regional, and national targets through evidence-based intervention.

As compared to urban women, rural women have been confronting with poor health service, since the many decades. This is mainly due to inadequate healthcare providers, especially women's health providers. This short of problem is more acute in the heterogeneous population locality, like America's rural areas. So the healthcare professional should be aware of the fact and advocate for reducing health disparities in rural areas.

Women confront with many unique life-threatening chronic diseases, besides the common chronic diseases for both men and women. Some exclusive health challenges being faced by women are pregnancy and menopause to gynecological conditions, uterine fibroids and pelvic floor disorders, cancers of the female reproductive tract, urinary tract health, bacterial vaginosis, and vulvodynia.

Rural women of both the developed countries and developing countries are victimization of chronic diseases. In the US, one of the leading causes of adult morbidity and mortality is chronic diseases like diabetes, cancer, and disease of cardiopulmonary system. Rural population in the United States have higher rates of chronic illness and poor health when compared to urban population. Very few rural adults have health insurance coverage due to high cost of insurance policy and lack of poor financial condition of senior citizens. More than 28 million women of rural America are badly in need of access to quality healthcare [60]. About 5 million adult women of age group 65 are residing in rural America out of which 4 million are identified with the problem of disability [60,61]. About 14% of rural adult women do not have health insurance [62]. From education point of view, women living in rural America are likely to have a high school diploma when compared to urban women as they are less likely to have college degree or higher [63]. Among the minority population in America, the majority of women are suffering from lower health status and higher rates of chronic illness. As per USDA statistics, rural America is home to more than 3.7 million women who are self-identified as African American, Hispanic/Latino, or American Indian [64]. All women in rural and frontier areas are affected by access issues, specifically the lack of primary and specialty care.

1.8.1 Aging issues

By 2050, the world's population aged 60 years and older is predicted to total 2 billion, up from 900 million in 2015. In 2017, the number of people having aged 80 years was about 125 million. By the end of 2050, the people aged 80 years would be about 125 million, as compared to 434 million people of same age group, worldwide [65].

Mainly, three demographic changes, i.e., (i) declining fertility, (ii) lower infant mortality, and (iii) increasing survival at older ages, are responsible for the increase in elderly aged population in the world. Currently, the two major challenges in demographic population distribution are (i) increase in the number of population having 80 years and above and (ii) feminization of aging as women live longer than men. It has been noticed that majority of elderly women population, about more than 55% of world population, live in developing countries. Within a short period of 5 years and so, nearly three-quarters of the world's older women are expected to reside in well-developed urbanized areas [66].

Most commonly, women tend to live longer than men. The ratio of women to men increases with age. Rural senior women are more likely to be disabled, widowed, older, and poorer than urban or suburban senior women. The elderly rural women lack access to many of the human services available to their urban and suburban counterparts. This can be directly correlated with well-being, independence, and quality of life of older rural women. For instance, senior women experience more health issues that affect their ability to drive than men. Due to lack of proper public transportation in rural areas, women face many hurdles to access primary healthcare center for treatment. In rural areas, women travel a long distance to reach a nearby primary healthcare center without the assistance of any elderly members from home, even during pregnancy; and responsibility to maintain daily livelihood for her own family and herself. Hence, the inability to drive can seriously hamper the mobility of rural elderly women, compromising their quality of life. The other challenges faced by older rural women include lonely life without her own family younger members, lack of knowledge of available services, and lack of needed services in or near the community. Under such unfavorable condition, the elderly rural women feel helpless to pull on with remaining life. Generally, rural elderly women suffer from old age health problems like macular degeneration, some types of cancer, and Alzheimer's diseases, very often. Generally, rural areas lack many social and health services to care for older women such as physicians specialized with gerontology and geriatrics and other specialists related to old age health problems [67,68]. In addition, elder women face difficulties to access in-home social services like adult day care, respite care, and meals on wheels.

During 2000–2050, the overall growth of population is expected to grow by 56%, while, the population 60+ will grow by 326%. At the same time, the population of 80+ will grow by 700% with a predominance of widowed and highly dependent elderly women. It has also been predicted that the number of older women will progressively increase as compared to older men. In 1999, the Government of India introduced National Policy on Older Persons (NPOP). So, the main target of NPOP is how to manage the aging population over the age of 60 and above in order to provide them proper health security for comfortable physical well-being. The National Social Assistance Programme for poor is also an outcome of the directive principles of our constitution (articles 41–42). Its main objectives are to develop social security and welfare program to provide support to aged persons, widows, disabled persons, and bereaved families belonging to below poverty line households.

1.8.2 Domestic violence

Domestic Violence is also known as domestic abuse or family violence. It is mainly related to domestic setting, such as marriage or cohabitation. In cohabitation, two unmarried people live together. They are often involved in a romantic or sexually intimate relationship for a longer or shorter period. Since the late 20th century, such arrangements become common in Western countries.

Traditionally, domestic violence used to be physical violence. Such violence includes wife abuse and wife beating. These short of unwanted family troubles come due to extramarital relation of husband or wife or sexual harassment. Sometimes economic status of the family brings family misunderstanding. Domestic violence is often used as a synonym for intimate partner violence. This is mainly happening when one of the individual involves in intimate relationship with other person in heterosexual or homosexual relationships or between former spouses or partners. Domestic violence is also applicable to violence against children, teenagers, parents, or the elderly. It may be in multiple forms, including physical, verbal, emotional, economic, religious, reproductive, and sexual abuse. Domestic murders include stoning, bride burning, and dowry death. Globally, women are more severely victimized due to domestic violence. They are also likelier than men to use intimate partner violence in self-defense. Domestic violence is among the most underreported crimes worldwide for both men and women [69]. Generally, due to mental preoccupancy of male attitude, the men are being overlooked by healthcare providers [70].

Awareness, perception, definition, and documentation of domestic violence differ widely from country to country. Domestic violence often happens in the context of forced or child marriage [71].

Domestic violence and child abuse are closely interlinked. The case of frequent domestic violence in a family badly affects the children and results in tremendous effect on their psychology and metal behavior. The estimated overlap between domestic violence and child abuse ranges from 30% to 50% [72].

Domestic violence against women is most commonly noticed in central Sub-Saharan Africa, eastern Sub-Saharan Africa, western Sub-Saharan Africa, Andean Latin America, northern Africa, and the Middle East. The lowest prevalence of domestic violence against women is found in Western Europe, East Asia, and North Africa.

1.8.3 Obstetric and reproductive health

Between 1990 and 2015, globally, maternal mortality ratio (MMR) has been significantly decreased by 44%, from 385 to 216 maternal deaths per 100,000 live births. In spite of such progress, the Millennium Development Goals (MDGs) didn't reach the scheduled target of 75% by the end of 2015. In 2015, MDGs were merged with Sustainable Development Goals (SDGs) to fulfil the target of 17 goals by the end of 2030. SDG3 is targeted for "Health for All" by the end of 2030. The World Health Organization (WHO) declared "Strategies toward ending preventable maternal mortality (EPMM)" for reducing maternal mortality at global level in collaboration with UNICEF. Globally, irrespective of urban and rural communities, little difference in obstetric outcomes is noticed [73].

Below, the problems of MMR in developed (America), developing (India), and least developed countries have been illustrated, briefly, to understand the severity of MMR, if serious efforts are not taken by the government. In addition, the strategy of SDG3 to help the developing and least developed countries to reduce MMR to a significance stage has also been explained.

The National Advisory Committee on Rural Health and Human Services, in its 87th submit, held on May 2020, examined maternal health and obstetric care challenges in rural America.

Mostly, rural women are prone to prenatal care due to limited availability of healthcare resources. In rural areas, obstetrics providers, in particular, are in short for looking after the pregnant cases, timely. The urbanized areas of developed countries like America, the number of obstetricians is about 35 per 1000 residence, while in rural areas of developing countries have less than 2 per 1000 residents [74]. Due to lack of such inadequate healthcare providers, the overall healthcare outcome in rural community is observed to be poor [75]. So, rural women rely upon Family Medicine Physicians to provide healthcare. With the increase in population, the provision for health services has been significantly declined [76]. Lower education, unplanned pregnancy, and inadequate transportation to a healthcare center have been associated with untimely prenatal care [77]. Many rural hospitals do not offer obstetrics, nor convey the shortage of nurses and trained staff [78,79]. Rural African American women avail more expensive insurance as compared to women with private insurance [80,81]. The non-white patients get less healthcare as compared to white women [82,83]. The non-white women are deprived of adequate access to family planning services that resulted in unplanned pregnancies. The decrease in unplanned family planning leads to fewer infants born with low birth weights, incidents of infant and neonatal deaths, and fewer abortions [84]. In order to facilitate the family planning program, many publicly funded agencies, such as community health center, public health department, hospitals, and women's health clinics almost universally offer contraceptive services to women [85,86].

In the US, more than 700 women a year die due to complications related to pregnancy, whereas two-thirds would have been preventable [87]. In 2016, the pregnancy-related mortality ratio was 16.9 per 100,000 live births [88]. This was mainly due to significant racial disparities (Fig. 1.19).

In America, the maternal mortality among American Indian and Alaska Native women is noticed to be more as compared to urbanized American. In 2015, the survey report from Centre for Disease Control and Prevention (CDCP) disclosed that about 29.4% per 100,000 live births occurred in rural areas, as compared to 18.2% per 100,000 in urban areas. Maternal mortality rate of rural black women in Georgia is 30% higher than the urban black women, and rural white women have a 50% higher risk than urban white women [89,90].

Severe maternal morbidity (SMM) like unexpected outcomes of labor resulted in short- or long-term consequences to a woman's health [91]. Since 2014, SMM has been increasing affecting more than 50,000 women in the US [91], and the risk of SMM in rural women is higher than the urban counterpart. This is mainly due to the lack of health workforce, social determinants of health like: transportation, housing, poverty, food security, racism, violence, and trauma.

In developed country like the US, the healthcare risk, mainly, depends on the location of rural area from urban territory. Prenatal care initiation in the first trimester is noticed to be lower for mothers in more rural areas as compared to urban locality [92].

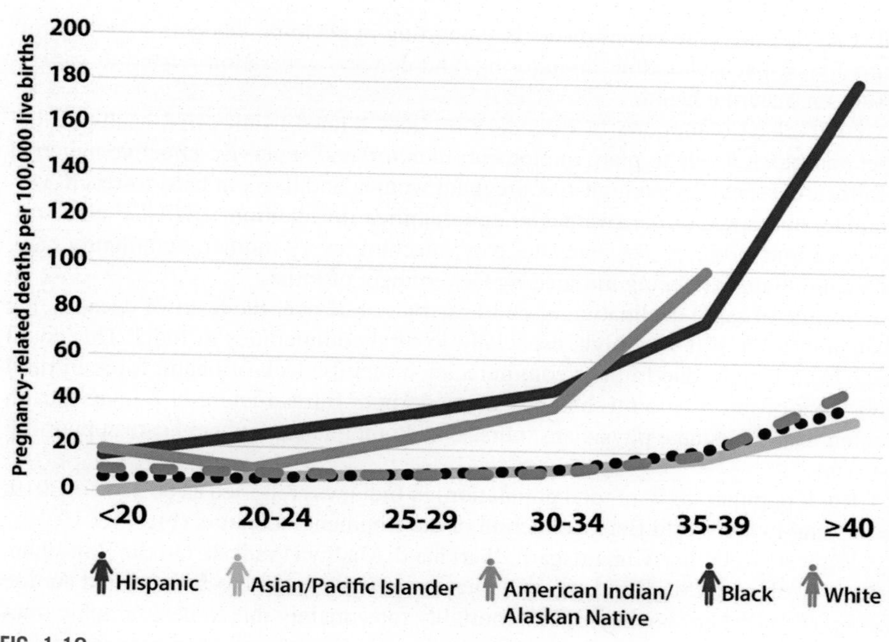

FIG. 1.19

Trends in pregnancy-related mortality ratios among race from 2007 to 2016.

Source: www.cdc.gov › reproductivehealth › maternal-mortality.

Generally, the rate of hospitality in rural area is comparatively higher than the urbanized locality [93].

The present (June 18, 2021) birth rate for India is 17.377 births per 1000 people, a 1.22% decline from 2020. The birth rate for India in 2020 was 17.592 births per 1000 people, a 1.2% decline from 2019 [94].

Two decades back, Maternal Mortality Ratio (MMR) in India was extremely high with 556 women dying during childbirth per 100,000 live births. This was mainly due to complications related to pregnancy and childbirth. As per the latest report from Sample Registration System (SRS) for the period 2016–2018, there was 113/100,000 live births, declining by 17 points, from 130/100,000 live births in 2014. This means 2500 additional mothers saved annually in 2018 as compared to 2016. Total estimated annual maternal death declined from 33,800 maternal deaths in 2016 to 26,437 deaths in 2018.

Currently, the Government of India is focusing on initiatives to improve maternal health indicators. Globally, the number of women and girls who die each year due to issues related to pregnancy and childbirth has dropped considerably, from 451,000 in 2000 to 295 in 2017, 138% decrease.

As per the target of SDG3, UNICEF in collaboration with Ministry of Health and Family Welfare (MoHFW), Ministry of Women and Child Development (MWCD),

NITI Aayog and state governments is supporting in planning, budgeting, policy formulation, capacity building, monitoring, and demand generation related to various issues on maternal health.

UNICEF also supports in extending services related to maternal health at district and block levels to plan, implement, monitor, and supervise effective maternal healthcare, especially on high-risk pregnant women and those in hard-to-reach, vulnerable, and socially disadvantaged communities. In addition, UNICEF also supports various projects like antenatal care, reaching every mother, community care, and continuum care being initiated by Government of India.

In spite of rapid declination of child mortality rate over the last two decades, the inequalities are still very high, especially in rural communities in India. This could have been mainly due to poor communication facility, lack of health force in rural area, lack of information technology network, poor infrastructure of primary health center, and inadequate physicians, nurses, and other auxiliary healthcare supporting workers.

Rural population (% of total population) in India was reported as 65.53% in 2019, according to the World Bank collection of development indicators (Fig. 1.20).

Eight states (Bihar, Chhattisgarh, Jharkhand, Madhya Pradesh, Orissa, Rajasthan, Uttaranchal, and Uttar Pradesh) of India are referred to as the Empowered Action Group (EAG) states having highest mortality rates mainly due to demographic transition [95]. In about 48% of India's population, EAG states are at greater risk due to the level of health disparities [96]. Despite sincere effort, even in association with UNICEF (as scheduled in SDG3), maternal and reproductive health is unacceptably emerging as a challenge in the backward states of India.

The highest attainable standard of health is a fundamental right of every person. Gender-based discrimination, however, undercuts this right. It can render

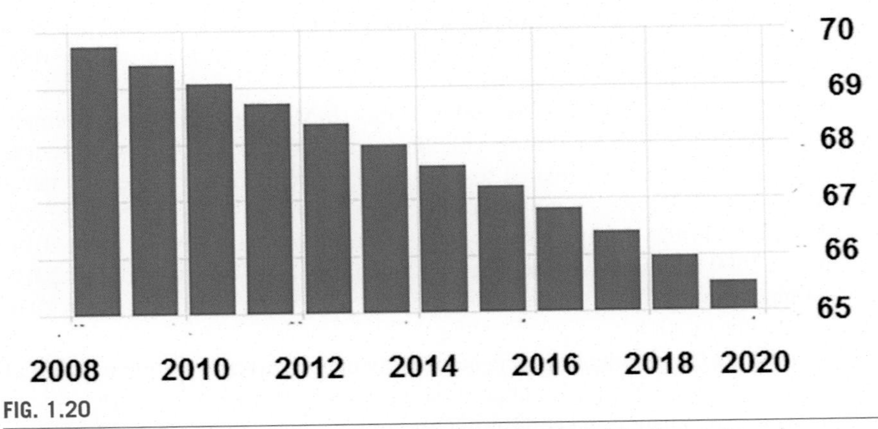

FIG. 1.20

Showing India—*Rural population—actual values, historical data, forecasts and projections were sourced from the World Bank on June of 2021. *Rural population refers to people living in rural areas as defined by national offices. It is calculated as the difference between total population and urban population.

women more susceptible to sickness and less likely to obtain care, for reasons ranging from affordability to social conventions keeping them at home. Among women of reproductive age worldwide, AIDS is now the leading cause of death. Not only are women biologically more susceptible to HIV transmission but their unequal social and economic status undercuts abilities to protect themselves and make empowered choices. Countries have committed to universal access to sexual and reproductive healthcare services, but many gaps have slowed progress so far. More than 225 million women have an unmet need for contraceptive methods. In developing regions, where maternal mortality rates are 14 times higher than in developed ones, only half of pregnant women receive the minimum standard for antenatal care. Fulfilling the right to health requires health systems to become fully responsive to women and girls, offering higher quality, more comprehensive and readily accessible services. Societies at large must end practices that critically endanger women.

References

[1] Ratcliffe M, Burd C, Holder K, Fields A. Defining Rural at the U.S. Census Bureau; American Community Survey and Geography Brief; 2016. p. 1–8.

[2] Economic and Social Council. Press release. Commission on the Status of Women to focus on rural women, their contributions - Challenges during fifty-sixth session at Headquarters, 27 February–9 March 2012 - Economic and Social Council; 24 February 2012. 24 February 2012, Press release.

[3] FAO The State of Food and Agriculture. Women in agriculture: closing the gender gap for development; 2010–2011.

[4] National Institute of Population Research and Training (NIPORT), Mitra and Associates, ICF International. Bangladesh demographic and health survey 2014: key indicators. Dhaka, Bangladesh and Rockville, MD: NIPORT, Mitra and Associates, and ICF International; 2015.

[5] Mishra V, Retherford R. The effect of antenatal care on professional assistance at delivery in rural India. Popul Res Policy Rev 2008;27(3):307–20.

[6] WHO., 2019, https://www.who.int/news-room/fact-sheets/detail/maternal-mortality.

[7] Efendi F, Ni'mah AR, Hadisuyatmana S, Kuswanto H, Lindayani L, Berliana SM. Determinants of facility-based childbirth in Indonesia. Sci World J 2019; [ID 9694602].

[8] ISGE. List of developing countries., 2018, https://isge2018.isgesociety.com/registration/list-of-de.

[9] UNCTAD. UN list of least developed countries, https://unctad.org/topic/least-developed-countries/list.

[10] McIlhenny C, Guzic B, Knee D, Demuth B, Roberts J. Using technology to deliver healthcare education to rural patients. Rural Remote Health 2011;11(4). www.rrh.org.au/journal/article/1798.

[11] Lustria ML, Smith SA, Hinnant CC. Exploring digital divides: an examination of eHealth technology use in health information seeking, communication and personal health information management in the USA. Health Informatics J 2011;17(3):224–43.

[12] Douthit N, Kiv S, Dwolatzky T, Biswas S. Exposing some important barriers to health care access in the rural USA. Public Health 2015;129(6):611–20.

[13] Boniol M, et al. Gender equity in the health workforce: analysis of 104 countries, health workforce working paper, no. 1. World Health Organization; 2019. http://apps.who.int/bookorders.
[14] Downs J, et al. Increasing women in leadership in global health. Acad Med 2014;89(8):1103–7.
[15] COVID-19 and its economic toll on women: The story behind the numbers., 2021, https://reliefweb.int/report/world/impact-covid-19.
[16] Stith AY, Nelson AR, Institute of Medicine, Committee on Understanding and Eliminating Racial and Ethnic Disparities in Health Care, Board on Health Policy, Institute of Medicine. Unequal treatment: confronting racial and ethnic disparities in health care. Washington, DC: National Academy Press; 2002.
[17] Heron M, et al. Deaths: final data for 2006. National vital statistics reports. Accessed at., April 17, 2009, http://www.cdc.gov/nchs/data/nvsr/nvsr57/nvsr57_14.pdf.
[18] Health, United States. 2014: with special feature on adults aged 55–64. Hyattsville, MD: NCHS; 2015. p. 53–5. http://www.cdc.gov/nchs/data/hus/2014/001.
[19] Martin JA, Hamilton BE, Osterman MJ, Curtin SC, Matthews TJ. Births: final data for 2013. Natl Vital Stat Rep 2015;64(1):1–65.
[20] Taylor P, Cohn D. A milestone en route to a majority minority nation. Washington, DC: Pew Research Center; 2012. Available at: http://www.pewsocialtrends.org/2012/11/07/a-milestone-en-route-to-a-majority-minority-nation.
[21] Anon, http://www.nimhd.nih.gov/documents/NIH%20Health%20Disparities%20Strategic%20Plan%20and%20Budget%202009-2013.pdf.
[22] Department of Health and Human Services. U.S. Department of Health and Human Services implementation guidance on data collection standards for race, ethnicity, sex, primary language, and disability status. Washington, DC: HHS; 2011. http://aspe.hhs.gov/basic-report/hhs-implementation-guidance-data-collection-standards-race-ethnicity-sex-primary-language-and-disability-status.
[23] UNICEF, UNFPA, World Bank Group and the United Nations Population Division, editors. Trends in maternal mortality: 2000 to 2017: estimates by WHO. Geneva: World Health Organization; 2019.
[24] Ganchimeg T, Ota E, Morisaki N, et al. Pregnancy and childbirth outcomes among adolescent mothers: a World Health Organization multicountry study. BJOG 2014;121(Suppl 1):40–8.
[25] Althabe F, Moore JL, Gibbons L, et al. Adverse maternal and perinatal outcomes in adolescent pregnancies: The Global Network's Maternal Newborn Health Registry study. Reprod Health 2015;12(Suppl 2):S8.
[26] Say L, Chou D, Gemmill A, Tunçalp Ö, Moller AB, Daniels JD, et al. Global causes of maternal death: a WHO systematic analysis. Lancet Glob Health 2014;2(6):e323–33.
[27] World Health Organization and United Nations Children's Fund. WHO/UNICEF joint database on SDG 3.1.2 Skilled Attendance at Birth. Available at: https://unstats.un.org/sdgs/indicators/database/.
[28] World Health Organization. Constitution of the World Health Organization - basic documents. 45th ed; 2006.
[29] Whitehead M. The concepts and principles of equity in health (PDF) (report). Copenhagen: WHO; 1990. p. 29. Reg. Off. Eur.
[30] World Health Organization. Women & health: today's evidence, tomorrow's agenda; 2009.
[31] McIlhenny C, Guzic B, Knee D, Demuth B, Roberts J. Using technology to deliver healthcare education to rural patients. Rural Remote Health 2011;11(4). www.rrh.org.au/journal/article/1798.

[32] Census Bureau US. New census data show differences between urban and rural populations. Washington, DC: U.S. Census Bureau; 2016. https://www.census.gov/newsroom/press-releases/2016/cb16-210.html.

[33] Singh GK, Siahpush M. Widening rural-urban disparities in all-cause mortality and mortality from major causes of death in the USA, 1969–2009. J Urban Health 2014;91(2):272–92.

[34] Meit M, Knudson A, Gilbert T, et al. The 2014 update of the rural-urban chartbook. Bethesda, MD: Rural Health Reform Policy Research Center; 2014.

[35] Blake KD, Moss JL, Gaysynsky A, Srinivasan S, Croyle RT. Making the case for investment in rural cancer control: an analysis of rural cancer incidence, mortality, and funding trends. Cancer Epidemiol Biomark Prev 2017;26(7):992–7.

[36] Singh GK, Daus GP, Allender M, et al. Social determinants of health in the United States: addressing major health inequality trends for the nation, 1935–2016. Int J MCH AIDS 2017;6(2):139–64.

[37] Zahnd WE, James AS, Jenkins WD, et al. Rural-urban differences in cancer incidence and trends in the United States. Cancer Epidemiol Biomark Prev 2017. https://doi.org/10.1158/1055-9965.EPI-17-0430.

[38] Doescher MP, Jackson JE. Trends in cervical and breast cancer screening practices among women in rural and urban areas of the United States. J Public Health Manag Pract 2009;15(3):200–9.

[39] Cole AM, Jackson JE, Doescher M. Urban-rural disparities in colorectal cancer screening: cross-sectional analysis of 1998–2005 data from the centers for disease Control's behavioral risk factor surveillance study. Cancer Med 2012;1(3):350–6.

[40] National Center for Health Statistics. Health, United States, 2012: with special feature on emergency care. Hyattsville, MD: NCHS; 2013. http://www.cdc.gov/nchs/data/hus/hus12.pdf.

[41] National Center for Health Statistics. Health, United States, 2011: with special feature on socioeconomic status and health. Hyattsville, MD: NCHS; 2012. http://www.cdc.gov/nchs/data/hus/hus11.pdf.

[41a] Benard VB, Coughlin SS, Thompson T, Richardson LC. Cervical cancer incidence in the United States by area of residence, 1998–2001. Obstet Gynecol 2007;110:681–6.

[42] Xiong X, Zhang Z, Ren J, Zhang J, Pan X, Zhang L, et al. Impact of universal medical insurance system on the accessibility of medical service supply and affordability of patients in China. PLoS One 2018;13(3). https://doi.org/10.1371/journal.pone.0193273, e0193273.

[43] Yu H. Universal health insurance coverage for 1.3 billion people: what accounts for China's success? Health Policy 2015;119(9):1145–52. https://doi.org/10.1016/j.healthpol.2015.07.008.

[44] Yip CM, Hsiao WC, Chen W, Hu S, Ma J, Maynard A. Early appraisal of China's huge and complex health-care reforms. Lancet 2012;379(9818):833–42. https://doi.org/10.1016/S0140-6736(11)61880-1.

[45] Xu L, Wang Y, Collins CD, Tang S. Urban health insurance reform and coverage in China using data from National Health Services Surveys in 1998 and 2003. BMC Health Serv Res 2007;7(1):1–14. https://doi.org/10.1186/1472-6963-7-37.

[46] Department of Agriculture, Economic Research Service. Population and migration: overview, http://www.ers.usda.gov/topics/rural-economy-population/population-migration.aspx.

[47] Department of Health and Human Services, Health Resources and Services Administration Maternal and Child Health Bureau. Women's health USA. Rockville, MD: DHHS; 2011. http://mchb.hrsa.gov/whusa11.

[48] Department of Health and Human Services, Health Resources and Services Administration Maternal and Child Health Bureau. Women's health USA. Rockville, MD: DHHS; 2011. http://mchb.hrsa.gov/whusa11.
[49] Department of Health and Human Services, Health Resources and Services Administration Maternal and Child Health Bureau. Women's health USA. Rockville, MD: DHHS; 2011. http://mchb.hrsa.gov/whusa11.
[50] National Center for Health Statistics. Health, United States, 2012: With special feature on emergency care. Hyattsville, MD: NCHS; 2013. http://www.cdc.gov/nchs/data/hus/hus12.pdf.
[51] Benard VB, Coughlin SS, Thompson T, Richardson LC. Cervical cancer incidence in the United States by area of residence, 1998–2001. Obstet Gynecol 2007;110:681–6.
[52] Bennett KJ, Olatosi B, Probst JC. Health disparities: a rural-urban chartbook. Columbia, SC: South Carolina Rural Health Research Center; 2008. Available at: http://rhr.sph.sc.edu/report/(73)%20Health%20Disparities%20A%20Rural%20Urban%20Chartbook%20-%20Distribution%20Copy.pdf.
[53] Hart LG, Larson EH, Lishner DM. Rural definitions for health policy and research. Am J Public Health 2005;95:1149–55.
[54] Rayburn WF, Richards ME, Elwell EC. Drive times to hospitals with perinatal care in the United States. Obstet Gynecol 2012;119:611–6.
[55] American Congress of Obstetricians and Gynecologists. The obstetrician-gynecologist distribution atlas. Washington, DC: ACOG; 2013.
[56] Koskinen S, Martelin T. Why are socioeconomic mortality differences smaller among women than among men. Soc Sci Med 1994;38:1385–96.
[57] Macran S, Clarke L, Sloggett A, Bethune A. Womens socioeconomic-status and self-assessed health-identifying some disadvantaged groups. Social Health Illness 1994;16:182–208.
[58] Arber S. Comparing inequalities in women's and men's health: Britain in the 1990s. Soc Sci Med 1997;44:773–87.
[59] Global, regional, and national comparative risk assessment of 79 behavioural, environmental and occupational, and metabolic risks or clusters of risks, 1990–2015. GBD 2015 risk factors collaborators. A systematic analysis for the global burden of disease study 2015. Lancet 2016;388(10053):1659–724.
[60] US Census Bureau. American community survey 5-year estimates, table DP05, 2007–2011. American FactFinder website http://factfinder2.census.gov.
[61] US Census Bureau. American community survey 3-year estimates, table B18101, 2009–2011. American FactFinder website http://factfinder2.census.gov.
[62] US Census Bureau. American community survey 3-year estimates, table B27001, 2009–2011. American FactFinder website http://factfinder2.census.gov.
[63] US Department of Health and Human Services, Health Resources and Services Administration. Women's Health USA, 2011. Maternal and Child Health Bureau website http://www.mchb.hrsa.gov/whusa11.
[64] US Department of Agriculture. Rural income, poverty, and welfare: poverty demographics. Economic Research Service website http://webarchives.cdlib.org/sw1rf5mh0k/, http://www.ers.usda.gov/Briefing/IncomePovertyWelfare/PovertyDemographics.htm.
[65] United Nations, Department of Economic and Social Affairs, Population Division. World population ageing 2017 - Highlights (ST/ESA/SER.A/397); 2017.
[66] Gist YJ, Velhoff VA. Gender and ageing: demographic dimension. Washington, DC: U.S. Bureau of Census; 1977.

[67] National Advisory Committee on Rural Health and Human Services. Report to the secretary: rural health and human service issues., 2010, http://www.hrsa.gov/advisorycommittees/rural/2010secretaryreport.pdf. Published May 2010.
[68] Fan L, Mohile S, Zhang N, Fiscella K, Noyes K. Self-reported cancer screening among elderly Medicare beneficiaries: a rural-urban comparison. J Rural Health 2012;28(3):312–9.
[69] Strong B, DeVault C, Cohen T. The marriage and family experience: intimate relationships in a changing society. Cengage Learning; 2010. p. 447, ISBN:978-1133597469.
[70] Finley L. Encyclopedia of domestic violence and abuse. ABC-CLIO; 2013. p. 163, ISBN:978-1610690010.
[71] WHO. Child marriages: 39,000 every day. who.int. World Health Organization; 2013.
[72] Rani M, Bonu S, Diop-Sidibé N. An empirical investigation of attitudes towards wife-beating among men and women in seven sub-Saharan African countries. Afr J Reprod Health 2004;8(3):116–36. https://doi.org/10.2307/3583398.
[73] National Center for Health Statistics. Health indicators warehouse, http://healthindicators.gov/Indicators.
[74] Health Resources and Services Administration. Area Resource File (ARF) 2011–2012. Rockville, MD: US Department of Health and Human Services, Health Resources and Services Administration; 2012.
[75] Nesbitt TS, Larson EH, Rosenblatt RA, Hart LG. Access to maternity care in rural Washington: its effect on neonatal outcomes and resource use. Am J Public Health 1997;87(1):85–90.
[76] Cohen D, Coco A. Declining trends in the provision of prenatal care visits by family physicians. Ann Fam Med 2009;7(2):128–33.
[77] Braveman P, Marchi K, Egerter S, Pearl M, Neuhaus J. Barriers to timely prenatal care among women with insurance: the importance of prepregnancy factors. Obstet Gynecol 2000;95(6):874–80.
[78] MacDowell M, Glasser M, Fitts M, Nielsen K, Hunsaker M. A national view of rural health workforce issues in the USA. Rural Remote Health 2010;10(3):1531.
[79] Zhao L. Why are fewer hospitals in the delivery business? Working paper 2007–04., April 2007, http://www.norc.org/PDFs/Publications/DecliningAccesstoHospitalbasedObstetricServicesinRuralCounties.pdf.
[80] Anum EA, Retchin SM, Garland SL, Strauss JFI. Medicaid and preterm birth and low birth weight: the last two decades. J Women's Health 2010;19(3):443–51.
[81] National Research Council (US) Panel on Race, Ethnicity, and Health in Later Life, Bulatao RA, Anderson NB. Understanding racial and ethnic differences in health in late life: a research agenda. Washington, DC: National Academies Press (US). 10, Health Care. Available from; 2004. https://www.ncbi.nlm.nih.gov/books/NBK24693/.
[82] Brandon GD, Adeniyi-Jones S, Kirkby S, et al. Are outcomes and care processes for preterm neonates influenced by health insurance status? Pediatrics 2009;124(1):122–7.
[83] Laditka SB, Laditka JN, Bennett KJ, Probst JC. Delivery complications associated with prenatal care access for Medicaid-insured mothers in rural and urban hospitals. J Rural Health 2005;21(2):158–66.
[84] Dobie SA, Gober L, Rosenblatt RA. Family planning service provision in rural areas: a survey in Washington state. Fam Plan Perspect 1998;30(3):139–47.
[85] Lindberg LD, Frost JJ, Sten C, Dailard C. The provision and funding of contraceptive services at publicly funded family planning agencies: 1995-2003. Perspect Sex Reprod Health 2006;38(1):37–45.

[86] Finer LB, Darroch JE, Frost JJ. U.S. agencies providing publicly funded contraceptive services in 1999. Perspect Sex Reprod Health 2002;34(1):15–24.

[87] Pregnancy-Related Deaths., 2019, https://www.cdc.gov/reproductivehealth/maternalinfanthealth/pregnancy-relatedmortality.htm.

[88] Pregnancy Mortality Surveillance System., 2019, https://www.cdc.gov/reproductivehealth/maternal-mortality/pregnancy-mortality-surveillance-system.htm.

[89] Warren, J. Maternal mortality in Georgia. http://www.house.ga.gov/Documents/CommitteeDocuments/2019/MaternalMortality/Mercer_University_Rural_Maternal_Health_Presentation.pdf.

[90] Severe Maternal Morbidity in the United States., 2017, November 27, https://www.cdc.gov/reproductivehealth/maternalinfanthealth/severematernalmorbidity.html.

[91] Kozhimannil KB, Interrante JD, Henning-Smith C, Admon LK. Rural-urban differences in severe maternal morbidity and mortality in the US, 2007-15., 2019, December, https://www.ncbi.nlm.nih.gov/pubmed/31794322.

[92] Agency for Healthcare Research and Quality. 2012 national healthcare disparities report. AHRQ Publication No. 13-0003. Rockville, MD: AHRQ; 2013. Available at http://www.ahrq.gov/research/findings/nhqrdr/nhdr12/nhdr12_prov.pdf. Retrieved October 30, 2013.

[93] Elixhauser A, Wier LM. Complicating conditions of pregnancy and childbirth, 2008. Statistical brief #113. Healthcare cost and utilization project (HCUP). Rockville, MD: Agency for Healthcare Research and Quality; 2011. http://www.hcup-us.ahrq.gov/reports/statbriefs/sb113.pdf.

[94] United Nations Department of Economic and Social Affairs: Population Division. World Population Prospects 2018. https//population,un,org>wpp.

[95] Ministry of Health and Family Welfare (Government of India). Health & population policies. NRHM; 2012.

[96] Kumar S, Sahu D. Socio-economic, demographic and environmental factors effects on under-five mortality in empowered action group states of India: an evidence from NFHS-4. Public Health Res 2019;9:23–9. https://doi.org/10.5923/j.phr.20190902.01.

CHAPTER

Rural women's health disparities, hunger, and poverty

2

2.1 Consequences of hunger and poverty on the health of rural women

Women are the backbone of the rural community and key actors in designing structural entities of rural livelihoods, being responsible for achieving nutritional security, generating income, and improving rural livelihoods and overall wellbeing. However, most often, rural women encounter poverty, food insecurity, and undernutrition, which result in poor health, within the limited four walls of a rural house (Fig. 2.1).

Rural women are active players in the progress of the Sustainable Development Goals (SDGs) established by the United Nations General Assembly in 2015 and intended to be achieved by the year 2030. The main challenges of the SDGs are to eradicate poverty and develop food security and nutritional quality for the health and wellbeing of the global population, especially with reference to rural communities where women are the worst sufferers. Food insecurity leads to poverty, which ultimately brings greater risks of chronic diseases and poor mental health. Beyond the consequences for individuals and families, these consequences also hamper economic progress and damage the health care system of a nation. Consequently, a group of international organizations have developed a task force on rural women, led by the Food Agriculture Organization (FAO), International Fund for Agricultural Development (IFAD), and World Food Programme (WFP), and composed of the following members: the International Training Centre of the International Labour Organization (ITCILO), Secretariat of the Permanent Forum on Indigenous Issues (SPFII), the United Nations Conference on Trade and Development (UNCTAD), United Nations Educational, Scientific, and Cultural Organization (UNESCO), the United Nations Population Fund (UNFPA), United Nations Industrial Development Organization (UNIDO), UN Women (United Nations Entity for Gender Equality and the Empowerment of Women), and the World Health Organization (WHO). The main goal of these organizations is to coordinate all forms of health promotion, poverty eradication, and community harmony, at a global level. In addition, these organizations also encourage and support research and development programs to improve food security, eradicate poverty, and to support economic stability.

Healthcare Strategies and Planning for Social Inclusion and Development. https://doi.org/10.1016/B978-0-323-90447-6.00006-0

FIG. 2.1

Showing how a rural family is managing their livelihood within the limited space of a rural house.

2.1.1 Strengthening rural women's livelihoods

Agricultural development is one of the most important factors to eradicate poverty at the global level and one which is unavoidable given our need to feed 9.7 billion people by 2050. However, despite the MDGs and SDGs programs, poverty reduction and food security are at risk. The food security and nutritional status of the most vulnerable population groups is expected to deteriorate further due to the health and socioeconomic impacts of the COVID-19 pandemic, according to the State Food Security and Nutrition in the World 2020 report [1]. This report also states that about 690 million people underwent hunger in 2019—up by 10 million from 2018, and by nearly 60 million in 5 years. Asia is the most effected by hunger, but it is expending fastest in Africa. Consequently, resolving food insecurity is one of the main agendas of the Sustainable Development Goals (SDGs) for the 21st century. Therefore, investment in agriculture and rural development to boost food production and nutrition is a priority for the World Bank Group. But without the improvement of rural women's status and empowerment, rural agricultural activities may not proceed well. As recently as a decade ago, more than a third of the female workforce was engaged in agriculture, while in regions like Sub-Saharan Africa and South Asia, more than 60% of all female employment is in this sector [2].

Rural women play a catalytic role towards achievement of transformational economics and the development of structure and function of rural communities around the world. However, due to inadequate facilities rural women have been facing poverty for a long time. This is further aggravated by the global food and economic

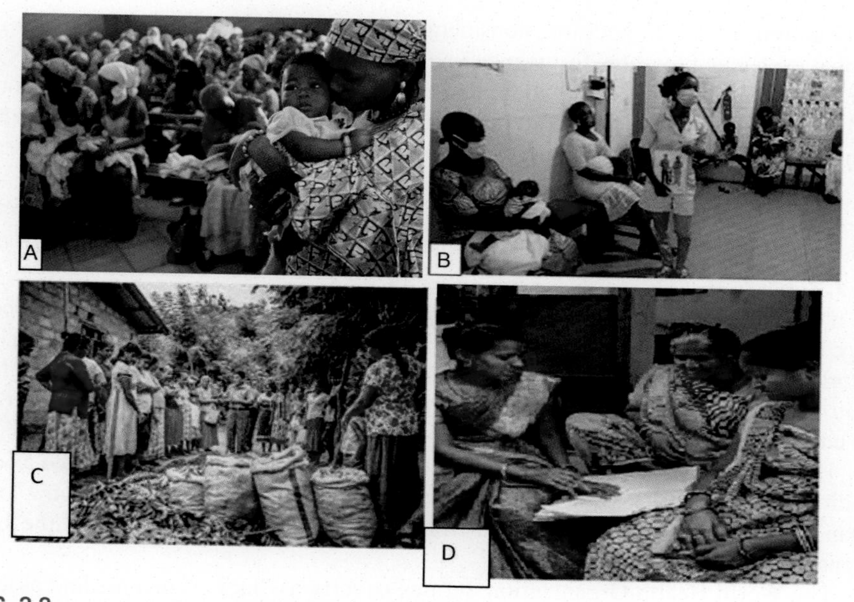

FIG. 2.2

Strengthening the rural women (A) by giving primary health education and (B) by providing health service.

crisis and climate change. Therefore, the overall economic development of a country depends on how we promote and empower rural women, and it is necessary to strengthen rural women by giving training on agricultural practices, providing them primary education and support for rural marketing and storing of food grains (Fig. 2.2).

October 15 is celebrated as United Nations' International Day of Rural Women. On this day, the campaign for recognition of rural women's importance in encouraging agricultural and rural development is highlighted. The International Day of Rural Women promotes awareness both of the contributions that women make in rural areas and the many challenges they encounter. The International Day of Rural Women was established by the UN General Assembly in its resolution 62/136 of December 2007, which recognizes "the critical roles and contribution of women, including indigenous women, in enhancing agricultural and rural development, improving food security and eradication rural poverty." The first International Day of Rural Women was observed on October 15, 2008.

The International Day of Rural Women is celebrated by many people, government agencies, community groups, and nongovernment organizations through television, radio, online, and print media broadcast publishing special features to promote the day. In addition many seminars, workshops, and conferences are held to review and update the status of rural women and their role in the community, particularly to promote the areas where there is a lack of economic development and agricultural output. In addition, consideration is given to women's involvement in global exchange programs for women in agriculture; launching fundraising projects for women in agriculture; exhibitions and workshops showcasing rural women's contribution to their societies; and strategic meeting to present issues on topics such as empowering women farmers to policy makers.

2.1.2 Poverty, health, and wellbeing

2.1.2.1 What is poverty?

In undeveloped countries of the world, poverty means living on less than $1.90 per day. This is a level set by the World Bank and is classified as "extreme" poverty. The World Bank data survey reports that about 700 million people (10% of world population) live below this threshold. They often have unsafe drinking water, have no access to education, and have a short life expectancy.

However, in developed countries like the United States, poverty is defined as an individual with income less than $36 per day or a family of four with income less than $72 per day. This is calculated from the poverty threshold as set by the US Census Bureau.

The World Health Organization defines poverty as "lack of income and assets to attain basic necessities—food, shelter, clothing, and acceptable levels of health and education, sense of voicelessness and powerlessness in the institutions of state and society. Vulnerability to adverse shocks, linked to an inability to cope with them."

According to the United Nations' statement, poverty is "more than the lack of income and resources to ensure a sustainable livelihood. Their manifestations include hunger and malnutrition, limited access to education and other basic services, social discrimination and exclusion as well as the lack of participation in decision-making. Economic growth must be inclusive to provide sustainable jobs and promote equality" [3].

Extreme poverty is a variable condition for individuals that varies over time due to changing prices of goods and services. In 2011, the person considered to be in extreme poverty was living with a $1.25 budget per day to survive, but in 2015, the World Bank announced that the extreme poverty threshold was $1.90 a day [4]. Extreme poverty, or absolute poverty, was originally defined by the United Nations in 1995 as "a condition characterised by severe deprivation of basic human needs, including food, safe drinking water, sanitation facilities, health, shelter, education and information" [5] (Fig. 2.3).

Mahatma Gandhi, a great thinker from India, said "Poverty is the worst form of violence" [6]. This statement he gave when people in Northern Kenya were struggling with poverty, poor nutrition, and higher risk of diseases; however, many Kenyan people still have lower life expectancy and inadequate access to healthcare. In 2005, a census by the Government of Kenya found that 91.7% of the individuals in Marsabit District were living below the national poverty line, making it the poorest of the 69 districts in Kenya (National Bureau of Statistics, 2012). Today, Kenya is a

FIG. 2.3

Showing how people are living under (A) extreme poverty and (B) absolute poverty.

lower-middle income economy. Although Kenya's economy is the largest and most developed in eastern and central Africa, 36.1% (2015/2016) of its population lives below the international poverty line [7]. This severe poverty is mainly caused by economic inequality, government corruption, and health problems.

With a population of 1.3 billion people, India is the world's largest democracy and the second largest country on the planet, after China. The COVID-19 pandemic struck India when India was confronting its lowest economic growth for over a decade. The slowing economy has disproportionately impacted the rural areas, where the majority of the country's consumers and poor reside. However, even in city life, people (mostly from the labor community) suffer from poverty (Fig. 2.4) and lack of shelter.

The Mahatma Gandhi National Rural Employment Guarantee Scheme is incapable of absorbing the demand for employment. The second wave of the pandemic aggravated the economic and health conditions, especially of rural people in the most difficult conditions. In effect, even the economic progress of the marginally well-off has ceased. As stated in the World Bank survey report, the number of poor in India (with income of $2 per day or even less) has more than doubled to 134 million from 60 million in just a year due to the pandemic-induced recession. In recent years, India emerged as the country with the highest rate of poverty reduction. In 2019,

FIG. 2.4

Daily laborers in Indian cities suffer from lack of shelter and sleep on roadsides.

the Global Multidimensional Poverty Index reported that India lifted 271 million citizens out of poverty between 2006 and 2016. India has not counted its poor since 2011. But the United Nations estimated the number of poor in the country to be 364 million in 2019, or 28% of the population. All the estimated new poor due to the pandemic is in addition to this.

2.1.2.2 Poverty and health

In general, poor people suffer the worst health and die at a younger age. Due to lack of a sound financial background, they have higher than average child and maternal mortality, higher levels of disease, more limited access to health care and social protection, and more severe gender inequality. As poverty and health are interlinked, it is necessary to simultaneously take care of both issues for rural community development (Fig. 2.5).

When elderly caregivers of a poor family, especially in rural areas, fall ill, the entire family and household suffer in managing daily livelihood. The household responsibility of women in rural areas is more critical than for male individuals. Today, rural woman shoulder all types of external and household work, ranging from caring for and schooling their children to taking care of elderly people and managing agricultural activities. Rural women who work in the agricultural field have to fulfill their job responsibilities while also managing home and family. The life of a rural woman is very tough, but she has been ignored and prevented from participating in leading society. There are many women who are extremely talented and possess multiple capabilities, but have no recognition in Indian society due to many social cultural restrictions.

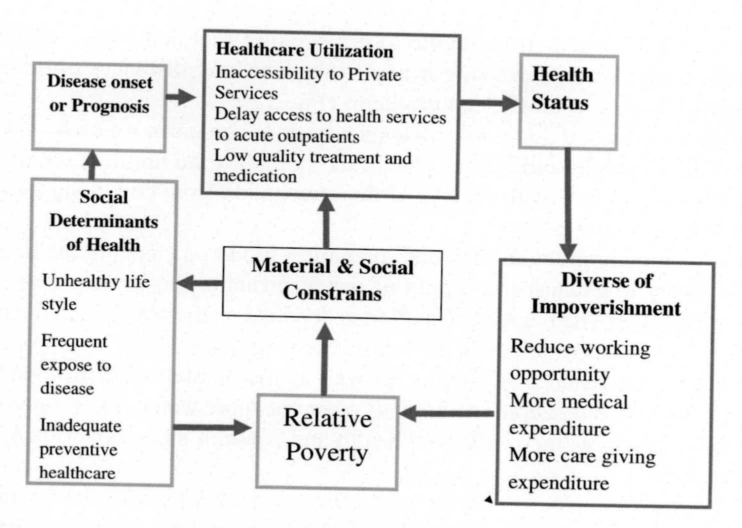

FIG. 2.5

Poverty-health interconnections for healthcare services.

The poor rural people are more vulnerable to disease, with limited access to healthcare and social insurance. Rural people are deprived of economic benefits (income, livelihood, decent work), primary health care, political influence (empowerment, rights, voice), social and cultural dignity, protection and security, and primary education, leading to social exclusion. Some social groups are particularly affected by severe poverty, among them indigenous populations, ethnic and minority groups, physically and mentally disabled individuals, and people living with HIV/AIDs.

The rural areas of both developed and developing countries have the problem of poverty and health, especially of rural women. According to the most recent estimates from the 2019 American Community Survey (ACS), the nonmetropolitan poverty rate was 15.4% in 2019, compared with 11.9% for metropolitan areas.

As compared to urban women, rural women spend more time in reproductive and household work. In addition, rural women may have to collect drinking water and fuel wood from local resources, sometimes at some distance from the village. This is mainly due to poor infrastructure and services as well as culturally assigned roles that severely limit women's participation in employment opportunities. For instance, collectively, women from Sub-Saharan Africa spend about 40 billion hours a year collecting water [8]. Due to nonavailability of conventional energy facilities and poor drinking water supply, rural women under great constrain carry enormous burdens of managing drinking water and fuel for their households. In rural areas of Guinea, for example, women spend more than twice as much time fetching wood and water per week than men, while in Malawi, they spend over eight times more time than men on the same tasks. Girls in rural Malawi also spend over three times more time than boys fetching wood and water.

Women in rural India still struggle to get cooking fuel and water, which hurts their health badly. Women carrying heavy loads over long distances often develop genital, spinal, and musculoskeletal problems (Fig. 2.6).

A bundle of firewood that a woman carries back to home can weigh 8–10kg. The quantity of firewood depends on the size of the family. If the family has cattle, then the load may be reduced with the use of dry cake made from cow dung as fuel for cooking (Fig. 2.7A and B).

After returning home with firewood, preparing food and cleaning the house and kitchen is likely to exhaust the supply of water, meaning the woman must go out again to fetch water. The number of such trips depends on the season and the number of family members and animals in the household (Fig. 2.8).

In summer, the distance she walks, as well as the number of trips, increases as hand pumps dry up, but people and animals consume more water. So, in summer, due to overload of work, women suffers ill health and is not in a position to manage the family properly.

Rural women confront a variety of hurdles in accessing affordable, adequate health services in rural communities (e.g., clinics, hospitals, reproductive health/ family planning, and counseling). This is partly due to the lack of affordable transport systems, restrictions on their mobility, and lack of communication networks. Rural women are at high risk of abuse, sexual harassment, and other forms of gender-based violence. This is mainly due to gender power imbalances, a lack of oversight, and working alone in relative isolation or in remote locations. Therefore, on a priority basis, rural women are in great need of holistic health services that can take overall care of the physical as well as mental and emotional wellbeing of rural women.

The World Health Organization, with the help of UNICEF, have many gender-based programs to resolve various health issues associated with rural life, with

FIG. 2.6

A woman carrying fire wood from distance forest.

FIG. 2.7

(A) A village women cooking by using cow dung. (B) Preparing cake from cow dung as fuel for cooking food.

special reference to rural women and girls. Other rural public health agencies also work to protect and improve the health of rural populations by:

- Preventing outbreaks/epidemics of communicable disease
- Providing emergency risk management in disaster conditions
- Promoting healthcare measures, especially in rural areas
- Providing emergency health relief during disasters
- Providing a guaranteed health service, especially during pandemic situations

FIG. 2.8
Rural women, with their children, carrying water to the village from a nearby resource.

The public health agencies face many challenges, like recruiting healthcare delivery persons, giving proper training to the healthcare workforce, making emergency arrangements for infrastructure for a telehealth network, and arranging funds.

Most of the international health agencies try to support SDGs goals of improved health and overall wellbeing for rural women through advocating for social protection, dedicated service, and elimination of child labor. The United Nations Population Fund (UNFPS) works on improving reproductive and maternal health worldwide. Its main targets are to provide adequate access to birth control, and leading campaigns against child marriage, gender-based violence, obstetric fistula, and female genital mutilation.

More than 150 countries across the four geographic regions (Arab States and Europe; Asia and the Pacific; Latin America and Caribbean; and Sub-Saharan Africa) receive support from UNFPS to mitigate the problems of hunger and poverty. It is a founding member of the United Nations Development Group, a collection of UN agencies that support Sustainable Development Goals.

In 1994, September 5–13, the United Nations coordinated an International Conference on Population and Development (ICPD) in Cairo, Egypt, with some 20,000 delegates from various governments, UN agencies, NGOs to discuss variety of population issues related to infant mortality, birth control, family planning, the education of women, and protection for women from unsafe abortion.

The latest ICPD was held in Nairobi in 2019. It was significant due to the completion of 25th year of the ICPD. The Governments of Kenya and Denmark and the UNFPA co-convened the summit, with a focus on pursuing sexual and reproductive health and rights (SRHR), which includes duty bearers, civil society organizations (CSOs), private sector organizations, etc., to discuss and agree on the actions to complete the ICPD Programme of Action.

The Programme of Action of the International Conference on Population and Development (ICPD) established UNFPA's mandate with an emphasis on (i) promoting sexual and reproductive health, (ii) upholding and expanding reproductive rights, and (iii) addressing population dynamics such as aging, migration, and the increase of population groups such as young people and adolescents. In each of these areas, UNFPA builds evidence and addresses the needs of poor rural women.

Still, rural culture does not allow women to claim their reproductive rights. In many countries, rural women are given less opportunity to share inherited wealth and property. In many rural communities, women and girls are vulnerable to the perpetuation of harmful practices such as child marriage, female genital mutilation (FGM), bride kidnapping, and widow abuse, and results of inadequate medical care, such as obstetric fistula.

Female genital mutilation (FGM) comprises all procedures that involve partial or total removal of the external female genitalia or other injury to the female genital organs for nonmedical reasons. There are strong protests from many countries in Africa, South East Asia, the Middle East, and from within communities from these areas, to stop this heinous practice (Fig. 2.9).

The WHO strongly urges health care providers not to perform FGM. It is recognized internationally as a violation of the human rights of girls and women. A half

FIG. 2.9

Anti-female genital mutilation (FGM) campaign near Kapochorwa, Uganda.

decade report from UNICEF disclosed that more than 200 million girls and women have been subjected to the practice, according to data from 30 countries where population data exist [9].

In 2007, UNFPA and UNICEF initiated the joint program on FGM to accelerate the abandonment of the practice. In continuation, in 2008, the WHO jointly with nine other United Nations partners published "Eliminating female genital mutilation: an interagency statement." In 2010, the WHO, in collaboration with other key UN agencies and international organizations, published a "Global strategy to stop health care providers from performing female genital mutilation." The WHO supports countries in implementing this strategy. In December 2012, the UN General Assembly adopted a resolution on the elimination of female genital mutilation. Finally, in 2018, the WHO launched a clinical handbook on FGM to improve knowledge, attitude, and skills of healthcare providers in preventing and managing the complications of FGM.

Considering that, as per the Sustainable Development Goals subclause SDG1, economic growth must be inclusive to provide sustainable jobs and promote equality, social protection systems, especially in rural areas, need to be implemented to help alleviate the suffering of disaster-prone countries and provide support in the face of great economic risks. In the short term this would be helpful to countries under the grip of the COVID-19 pandemic, and will eventually end extreme poverty in the most backward areas.

2.1.3 Social determinates and rural health

On May 07, 2013, the World Health Organization defined social determinates as "The circumstances, in which people are born, grow up, live, work and age, and the Systems put in place to deal with illness. These circumstances are in turn shaped by a wider set of forces: economics, social policies and politics."

The status of health is mainly dependent on income-level, educational attainment, race/ethnicity, and health literacy (Fig. 2.10).

All these facts directly impact on the ability of people to access health services and to meet their basic needs, such as clean water and safe shelter, to lead a sustainable healthy life. This can be challenged to eradicate poverty from rural communities. But the impact of these challenges can be compounded by the barriers already present in rural areas, such as limited public transportation options and fewer choices to acquire healthy food.

The social determinants of health (SDH) are the nonmedical factors that influence health outcomes. They are part and parcel of the circumstances in which individuals are born, grow, work, live, and age. But wider ranges of forces, like economic policies and systems, development agendas, social norms, social discrimination and disparities, and economic policies and systems, are major barriers to implementation of various programs related to rural health services. The followings are a few important barriers responsible for economic instability and rural health conditions:

- Per capita daily/monthly income and level of poverty
- Primary health education

FIG. 2.10

Village women fetching fuel wood from forest.

- Race and ethnicity
- Educational attainment and literacy
- Sexual orientation/gender identity
- Environmental health including water quality, air quality, and pollution
- Access to safe and healthy homes, including issues related to energy costs and weatherization needs
- Adequate community infrastructures, which can ensure public safety, allow access to media, and promote wellness
- Access to safe and affordable transportation, which can impact both job access and healthcare access

2.1.3.1 Economic stability and rural women

Economic stability and rural women are complementary with each other. Both the women from developed countries and developing countries often suffer from economic crises, which is ultimately responsible for poor health. In the United States, one in 10 people, including women, live in poverty. Many people are incapable of affording healthy food, healthcare, and housing. Healthy People 2030 focuses on helping more people achieve economic stability.

In 1979, the United States Department of Health and Human Services set "Healthy People" goals for health promotion and disease prevention. This program was subsequently updated for Healthy People 2000, Healthy People 2010, Healthy People 2020, and the future program, Healthy People 2030 [10]. The initiative related to healthy people began in 1979, when Surgeon General Julius Richmond issued a landmark report entitled, "Healthy People." It was mainly on various issues related to health promotion and diseases prevention. In September 1990, the Department of Health and Human Services released Healthy People 2000 with the motto to improve health of Americans. Healthy People 2000 contains 319 unduplicated main objectives grouped into 22 priority areas. This program was subsequently updated for Healthy People 2000, Healthy People 2010, Healthy People 2020, and the future program, Healthy People 2030 [10]. The first issue related to healthy people contained ten-year plan to reduce controllable health risks, and recommended low calories diet, less saturated fat, cholesterol, salt, and more complex carbohydrates, fish and poultry products, and less red meat [11]. The following are a few important factors responsible for health stability under various socioeconomic circumstances.

Generally, people with steady employment are less prone to poverty and may be supposed to lead a healthy life, but finding a regular job is difficult. Furthermore, despite remaining in a regular job, individuals may suffer poverty due to some other social or family reasons. People with disabilities, injuries, or conditions like arthritis may be especially limited in their ability to work, and consequently suffer poverty. Employment programs, career counseling, and quality childcare opportunities can help more people find and keep jobs. In addition, government policies to provide subsidy or easy installment-based housing loans, free education at primary level, and provision of affordable healthcare insurance may be helpful for a poverty-free, healthy life.

In developing countries, women play key roles in the rural economy as farmers, wage earners, and entrepreneurs. But, they are still legging beyond their urban counterparts, especially in relation to access to health services.

Women play an integral part in farming, either as a principal operator or as a decision-maker. About 41 % of the world's agriculture labor force is from rural sector of developing countries. However, in South-East Asian and Sub-Saharan African countries, the proportion of women in the workforce is about 60%. Most of these rural women manage their own small parcels of land, which require a great deal of work for marginal benefit; and under unfavorable conditions they are extremely vulnerable to a lack of basic investment. Besides, agriculture, in some countries rural women work in education, tourism, and domestic work. It has also been noticed that in Latin America, women are increasingly involved in nonagricultural sectors or other occupations, which give some financial return [12]. But in some countries, due to migration of the male population to cities for better income, the women play a significant role in compensating for the lost work of those men in agricultural activities [13]. In spite of their crucial roles in the rural economy, women face inequalities and challenges to their access to decent work opportunities and improvements in their productivity. About 68% of women working in the rural agriculture sector, including fisheries, forestry, handicrafts, and livestock rearing, are under extreme poverty [14]. Other challenges that women face in the rural economy include lack of information on job availability, as well as opportunities for training and education, limited access to property, land, and financial and nonfinancial services [15].

Generally, rural women remain busy in multiple activities while managing the family livelihood. As compared to urban women, rural women spend more time in reproductive and household work, including time for collecting firewood and drinking water, collecting green livestock (fodder) for cattle (Fig. 2.10) and animals as food, preparing food, and managing the entire household work.

Women's contribution to rural agriculture development and food security is significantly higher than that of men, but they are paid 22%–29% less than their male counterparts [16,17]. The wage disparity in rural women has been a continuous chronic factor. In addition, they are engaged in many occupational activities without any safety and healthcare measures, and social protection [18], including protection from sexual and other violence and harassment [10,19]. Globally, out of the total daily labor workforce, 71% are victims of forced labor, including 28.7 million women [20]. Most women in forced labor are found in agriculture, forestry, and fishing.

Migration from rural to urban areas has been a serious problem effecting rural economies. This problem is not a new one. Since colonial times, some groups of West African people have migrated for a better livelihood. West Africa has experienced a variety of migration due to population pressure, poor economic performance, and endemic conflicts. Migrants from West Africa have included temporary cross-border workers, traders, farm laborers including women, professionals, and refugees. This migration is mainly an intraregional and short-term response to the interdependent economics of neighboring countries [21].

Many people from underdeveloped African countries migrate to urbanized places of Europe, in hopes of a better life. In parts of Africa, particularly North Africa (Morocco, Mauritania, and Libya), trafficking immigrants to Europe has become more lucrative as better source of earning [22–24]. The impact of male migration in Africa, particularly in West Africa, has been well documented [25,26]; whereas documentation on female migration in West Africa is poorly available. Mainly, the factors responsible for female migration in West African countries are education, labor, marriage, trading, etc. Female migrants are also involved in the wage labor market (both formal and informal) as a better source of family income. Emigration of educated unaccompanied females has been steadily increasing. Migration of uneducated women trends to be more a common phenomenon of married women accompanying their husbands. Migration of rural women to urban areas has many benefits that can help them achieve a better livelihood, social empowerment, such as independence, economic attainment through trading, and networking to find better economic opportunities.

Initiation of urbanization in India began after independence. People from villages preferred to migrate for better livelihoods and healthcare. This was mainly due to the adaptation of a mixed economy, which resulted in development of the private sector. In 1901, the population residing in urbanized areas was about 11.4% of total population [27]; in 2001, this had increased to 28.53%, and in 2020, according to The World Bank, it was 34.53%. According to World Bank and a survey by the UN, in 2030, 40.76% of the country's population is expected to reside in urban areas [28]. As per the World Bank, India, along with China, Indonesia, Nigeria, and the United States, will lead the world's population surge by 2050 [29]. In 2018, Mumbai accommodated 22.1 million people, and is the largest population in India, followed by Delhi with 28 million inhabitants.

Rapid urbanization in India has resulted in problems like increasing slums, decreasing standard of living in urban areas, and deterioration of environmental conditions [30]. The Industrial Revolution of the 18th century made the United States and United Kingdom superpowers, but conditions elsewhere are worsening.

The urbanization process in India leads to issues like overcrowded cities, many people are forced to live in unsafe conditions such as illegal buildings, poor or no-navailability of water lines, and unsafe electricity supply, which has resulted in a decline in living standards (Fig. 2.11A–C).

Over the last few decades, there has been increasing feminization of international labor migration in the most of the developing countries [31–39]. The most common reason for internal migration in India is marriage. According to the 2011 Census report, 46% of the total migrants moved because of marriage, out of which 97% were women. Although marriage is the primary cause of migration, there has been an in increase in migration for economic reasons [40–44]. In India, a few small-scale surveys have been conducted to understand the basic reason of male migration, but no information is available on internal female migration in India [45,46].

Migration is a highly gendered phenomenon. Women, in general, face different challenges for migration due to social norms, lack of access to infrastructure and communication facilities, and whether they decide to migrate or stay behind as family of male migrants. Migration is out of reach for many rural women. Generally, rural women face problems to access resources (e.g., information, land ownership, assets, and social networks) for migration. Even where these resources are available, the larger family may control them, constraining women's migration opportunities. In addition, rural women are more prone to physical disorders and restrict their movement due to reproductive responsibilities. Historically, cultural norms and social exclusion have also restrict rural women's migration. Women who live in extreme rural localities often lack proper identity documents and find it difficult to access transport or information services.

Rural women may desire to lead a life where traditional gender roles, gender disparity, and gender-based violence do not exist. So, in some rural localities, women prefer to migrate to an urban life—after having transferred their respective family properties—in order to have a better, more trouble-free life. The best option for migration from a rural locality to an urbanized area is entire-family migration, as global demand for labor rises in highly gendered niches such as domestic work, healthcare for children and the elderly, and also in the garment and other small industries. This demand acts as a powerful "pull factor" for women in depressed rural areas. But, even in urbanized life, the migrated rural women face risks like abuse (those who work in poorly regulated sector), discrimination, and exploitation. This is mainly due to vulnerabilities as women, as outsiders, and being of rural origin. In spite of

FIG. 2.11

Views of unauthorized colonies with lack of water supply, electricity, and other public health facilities in (A) along railway track and in (B) along sewerage line.

such problems, the better remittances they get as compared to village laborers may enable them to finance their family members living in their respective native rural areas. This would be immensely beneficial to improve the standard of living at family level, and in the longrun, for the development of rural communities and eradication of poverty and promotion of health.

Healthcare access and quality

The Sustainable Development Goal 3 is targeted to ensure healthy lives and promote wellbeing for all ages. The main emphasis of this SDG is to achieve universal health coverage (UHC) and access to quality health care by the end of 2030. Nevertheless, rural inhabitants still confront a variety of access barriers. Access to healthcare means having available "the timely use of personal health services to achieve the best health outcomes" [47].

In rural areas, due to lack of healthcare providers, residents need to use unconventional ways of accessing medical consultations by phone or internet, as well as mobile preventive care and treatment programs. Presently, due to shortage of physicians to work in rural areas, various government and nongovernment organization have engaged nonphysician providers to deliver health care in rural areas.

Access to health care may vary across countries and individuals, and be influenced by social and economic conditions as well as health policies. Although access to healthcare services is critical to good health, still, rural people face a lot of hurdles to access healthcare services. In 2014, the RUPRI (Rural Policy Research Institute) Health Panel reviewed the various issues on rural healthcare access. The review is a highly informative piece of work on the paradigm of health strategies that would be immensely beneficial to researchers and policy makers. Access to rural healthcare has been defined in many ways by variousl researchers [48–53]. The objectives for access to rural health are significant for:

- Maintenance of physical, social, and mental health status, irrespective of race and ethnic demarcation
- Disease prevention, especially contagious disease
- Primary healthcare for detection, diagnosis, and treatment of health disorders
- To promote quality of life
- To reduce premature morbidity and mortality
- Old age healthcare services

Rural inhabitants often confront a variety of hurdles while trying to access healthcare services. Even in the case of available healthcare services, rural residents may be unable to pay nominal service charges, like payment of installments for life insurance. So, to avail themselves of the opportunity for quality health services, a rural resident must possess:

- The ability to pay installment fees for health insurance or fees to health service providers
- The ability to meet transport charge to visit nearby primary healthcare centers
- Self-confidence in their ability to communicate with healthcare providers
- Trust in healthcare providers without compromising privacy

Inadequate healthcare access and rural community health

At a global level, rural residents have experienced inadequate access to health services in terms of quality, affordability, and proximity, as compared to nonfarm and urban counterparts. Mostly, in developing country the primary health care centers are located far away from rural locality. Rural residents get limited access to healthcare, due to inadequate transport system and road ways. This problem can be resolve by providing well trained male and female primary healthcare providers to immediately attend the patients, before making proper arrangement to send them to nearby health center. The challenge that rural residents face is in accessing healthcare services, which contributes to health disparities. Barriers to healthcare result in unmet health-care needs, including a lack of preventive and screening services and treatment of illnesses. A vital rural community is dependent on the health of its population. While access to medical care does not guarantee good health, access to healthcare is critical for a population's wellbeing and optimal health.

Healthcare infrastructure problems in northern and rural areas of Canada mean that health disparities are greater than in urban areas. In Canada, rural residents have access to half as many physicians (one per 1000 residents) as their urban counterparts. As a result, people in northern and rural regions typically travel greater distances to obtain services that are available in their local communities. To avail themselves of health services, a rural patient has to travel about 10 km to access these services [16]. The primary rural health centers are poorly equipped with healthcare facilities and have a restricted number of specialists and nurses [54].

The gap in services is mainly due to allocation of more funds to urbanized areas as compared to rural areas. Due to a wider range of economic standards, the healthcare system in China has been rated as 144th in the world by the World Health Organization (WHO). The country spent only 5.6% of its GDP on health and has a relatively low number of doctors (1.6 per 1000 people).

Presently, Chinese people have three medical insurance systems available: (i) Urban Residents Basic Medical Insurance, (ii) Urban Employee Basic Medical Insurance, and (iii) the New Rural Co-operative Medical Scheme. These medical schemes cover almost everyone, however, China has large health disparities in rural areas as compared to urban localities. Rural patients in China face significant hurdles to accessing healthcare access. There are less than half as many medical institution beds and licensed physicians per 1000 citizens in rural areas compared to urban areas (Fig. 2.12) [55].

The Healthy China 2020 project is quite an ambitious program, under which the Chinese government was working on providing affordable basic healthcare to all residents by 2020. In 2018, the Chinese Government cut healthcare costs, requiring insurance to cover 70% of total costs. China has also become a major market for health-related multinational companies like AstraZeneca, GlaxoSmithKline, Eli Lilly, and Merck. So China has been becoming a hub for multinational healthcare companies.

Similar to Japan, healthcare in South Korea is of a very high standard, with about two doctors per 1000 people. The South Korean healthcare system is ranked 58th in the world by the WHO. South Korea has a universal healthcare system, within which a significant portion of healthcare is privately funded. Modern and efficient, both Western and traditional Eastern medicine is covered by the government's health

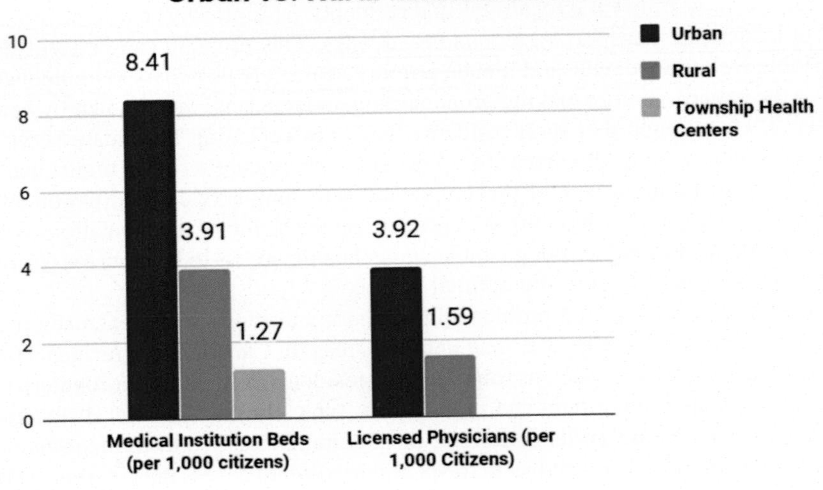

FIG. 2.12

Differences in urban and rural area healthcare infrastructure.

Adapted from the National Bureau of Statistics. Source: Collective Responsibility.

insurance scheme. From a healthcare disparity point of view, there is hardly any different between rural and urbanized areas.

In Japan, access to healthcare system is universal. This system is available to all citizens, as well as non-Japanese citizens staying in Japan for more than a year. Japan's Statutory Health Insurance System (SHIS) covers 98.3% of the population, while the separate Public Social Assistance Program, for impoverished people, covers the remaining 1.7%. Japan's government covers healthcare services, including screening examinations, prenatal care, and infectious disease control, with the patients accepting responsibility for 30% of these costs while the government pays the remaining 70%. All residents of Japan are required by law to have health insurance coverage. Local government provides health insurance programs for those people who do not have insurance from their employers. In Japan, patients have open option to select their physician or the facilities of their choice, but without insurance coverage. Hospitals, by law, must be run as nonprofit and be managed by physicians. Medical fees are strictly regulated by the government to keep them affordable. On the basis of age and the nature of services, the patient pays 10%–30% of the medical fee, and the remaining amount is paid by the government. In addition, the monthly thresholds are set for each household according to their income and age, and medical fees exceeding the threshold are waived or reimbursed by the government. Uninsured patients pay 100% of their medical fees, but patients belonging to low-income households receive a government subsidy.

In 2020, report on healthcare services, India ranked 112 (four places down from the previous year). India is lagging behind neighboring countries (China, Nepal,

Sri Lanka, and Bangladesh) in access to healthcare services. India has a universal multipayer healthcare model that is paid for by a combination of public and private health insurance along with the element of almost entirely tax-funded public hospitals. Nevertheless, rural India suffers a lot in connection with access to healthcare services. Rural areas in India are extremely short of medical professionals [56]. The doctor-population ratio in India is 1:1456 against the WHO recommendation of 1:1000. The survey also mentioned the initiatives the government has taken to address the shortage of doctors. At present, according to Union Health Ministry India data, there is an overall shortfall of 76.1% in specialist doctors at the Community Health Centers (CHCs) in rural areas of India as compared to the requirement for existing CHCs [57].

In spite of being one of the most populous countries, India has the most private healthcare in the world [57]. About 75% of total expenditure is being paid from the pockets of individual citizens [58,59]. Only one fifth of healthcare is financed publicly. Overall, India's public health expenditure (sum of central and state spending) has remained between 1.2% and 1.6% of GDP between 2008 and 09 and 2019–20. This expenditure is relatively low as compared to other countries such as China (3.2%), the United States (8.5%), and Germany (9.4%).

Switzerland's healthcare is universal and is regulated by the Swiss Federal Law on Health Insurance. Health insurance is required for all persons living in Switzerland. Swiss healthcare is not tax-based or financed by employers. Swiss health schemes are responsible for collecting nontax-based health insurance. Health insurance covers the costs of medical treatment and hospitalization of the insured. Swiss healthcare schemes are responsible for collecting individual payments on an installment basis. The basic health insurance coverage covers 80%–90% of healthcare costs, including outpatient treatment, emergency treatment, prescriptions, maternal medicine, vaccinations, and postoperational rehabilitation. In Switzerland, patients are never kept waiting for physicians as the government merged private, subsidized private, and public healthcare systems to provide its citizens with an extensive network of qualified doctors, and the best-equipped medical facilities and hospitals.

Finland is a high-income country located in northern Europe with a population of 5.5 million. It has a GDP of about €40,000 per person, the high standard of social and living conditions typical of a Nordic welfare state, and a low poverty rate. Finland's healthcare system is believed to be one of the best in the world. Finland offers its residents universal healthcare. Finland's citizens have access to health services, regardless of their financial situation. Public health services are mainly financed from tax revenues—partly municipal, partly state tax. The municipal authorities receive contributions for healthcare services from central government on the basis of population numbers, age structure, and morbidity survey reports. In Finland, 7% of its gross national product is spent on healthcare services. 76% of total health expenditure is funded by the public sector and the remaining amount by users of services and other sources (employers, private insurance, and benefit societies).

2.1.3.2 Education access and quality

Education is a fundamental human right and an enabling right. Therefore, it is necessary to ensure universal equal access to inclusive and equitable quality education and learning, which should be free and compulsory, for each and every person. The basic aim of education is to promote human personality and understanding, tolerance, friendship, and peace. Sustainable Development Goal 4 (SDG 4) is targeted to "Ensure inclusive and equitable quality education and promote lifelong learning opportunities for all" (Fig. 2.13). It is one of the 17 SDGs established by the United Nations in September 2015.

SDG4 is assessed by 10 indicators (Fig. 2.14), out of which seven are "outcome-oriented targets," and include:

Free primary and secondary education

In spite of good progress, achieving the Millennium Development Goal of universal primary education by 2015 was not completed as expected. In 2013, about 59 million children of primary school age, typically between 6 and 11 years, were out of school. As reported by the UNESCO Institute for Statistics (UIS), one out of five out-of-school children will never set foot in a primary school. In addition, 65 million adolescents of lower secondary school age, typically between 12 and 15 years of age, were not in school in 2013. Many are poor and live in rural areas, and suffer discrimination because of ethnic origin, language, gender, or disability. So keeping in view the above facts and figures, the target SDG 4.1 is aimed to provide all girls and boys equitable and quality primary and secondary education by the end of 2030. In addition, it is also targeted to ensure 12 years of free quality education funded by the public sector, of which 9 years are compulsory, leading to a relevant learning outcome.

FIG. 2.13

Symbol for Sustainable Development Goal 4.

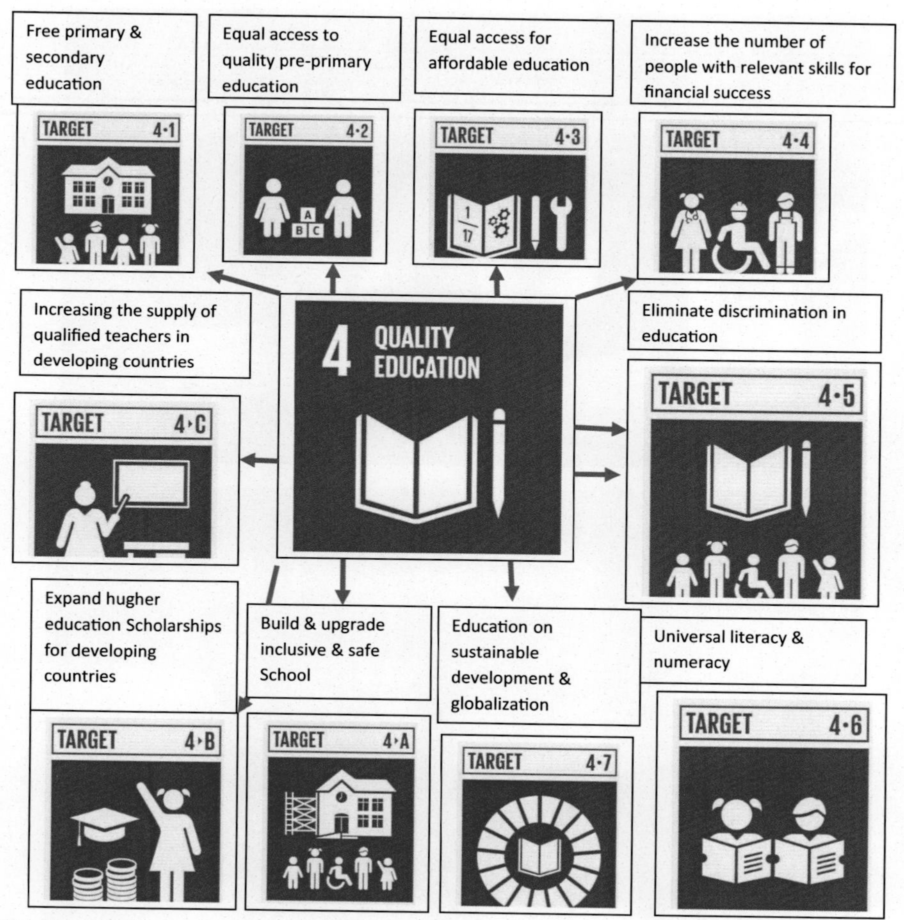

FIG. 2.14

Various aspects of SDG4.

Equal accesses to quality preprimary education

Early childhood development should be accompanied by the physical, social, emotional, and mental conditions to make children more independent, and allow them to learn increasingly advanced skills capacities as they grow older, This mainly depends on the surroundings in which they live and parental and social backgrounds, such as ethnic or religious groups within the same country (Fig. 2.15).

Irrespective of the pace and rate at which children develop, all children must have an equal right to develop to their fullest potential. Generally, children of ages three and four are in the development stage. Early childhood development, care, and education should be holistic, especially for girls residing in rural areas. The provision of at least 1 year of free and compulsory quality preprimary education is encouraged,

FIG. 2.15

Children learning in different environmental conditions. (A) Children reading in urbanized school and (B) children reading in village under tree due to lack of school infrastructure.

to be delivered by well-trained educators. However, this is heavily dependent on the status of development, availability of resources, and infrastructure of a country. Early childhood care is helpful in identification of disabilities and children at risk of disability, and accordingly, the healthcare providers or educators can plan a better way to overcome the problems in a timely manner. The United Nations Educational,

Scientific, and Cultural Organization (UNESCO), the United Nations International Children's Emergency Fund, and the World Bank are working with partners to enable members states to develop and use appropriate metrics.

So, the object of SDG4.2 is to ensure that all girls and boys must be provided the facility for early childhood development, and compulsory quality preprimary education by well-trained educators, so that the children will be fit for primary education.

Affordable technical, vocational, and higher education

To ensure the barriers to skills development and technical and vocational education and training (TVET) covering both the secondary level and tertiary education, including university, and to provide lifelong learning opportunities for youths and adults, a number of countries have taken steps to expand vocational education to the tertiary education level. Yet a wide disparity in access to tertiary education, in particular at university level, with regard to gender, social, regional, and ethnic background, and to age and disability, remains. Disadvantages for females occur particularly in low-income countries, and for males in high-income countries.

Increased number of people with relevant skill for financial success

The main target is to promote both youths and adults having relevant skills in technical and vocational knowledge for employment, decent jobs, and entrepreneurship. Adequate access should be provided to youths and adults, especially girls and rural women, to acquire relevant knowledge, skills, and competencies for decent work and life. Furthermore, in empowering the youth with technical skills, it is also necessary to develop self-confidence and communication skills, which can be helpful in the occupational field.

Elimination of discrimination in education

Gender disparities have been a chronic problem in education in developing countries. The SDG5 is targeted to ensure equal access to all level of education and vocational training for the vulnerable, including persons with disabilities, indigenous peoples, and children in vulnerable situations.

By the end of 2030, all people, irrespective of cast, sex, race, ethnicity, religion, language, and other social exclusive factors should have equity of opportunity in access to education systems. Both women and men should have equal levels of and enjoy equal benefits from education. Special attention must be paid to adolescent girls and young women, who may be subject to gender-based violence, child marriage, early pregnancy, and a heavy load of household chores, as well as residing in poor and remote rural localities.

Universal literacy and numeracy

The main target is to ensure that all youths and a substantial proportion of adults, both women and men, achieve literacy and numeracy, at a global level. Literacy helps students apply reading, writing, and speaking across a variety of subject areas. Numeracy is the ability to understand and apply mathematical concepts, process, and skills to solve problems and make decisions in a variety of situations, including real-life.

To understand the functional reality of adult literacy, first, the global indicator on literacy and numeracy is to be formulated explicitly in terms of skills proficiency. This will bring the understanding that literacy as not just a set of basic cognitive skills but also the ability to use them to contribute to societies, economies, and for personal change. Second, explicitly referring to numeracy calls attention to its properties.

So, the basic understanding of literacy is not just "literate" vs "illiterate." Therefore, action for this target aims at ensuring that, by 2030, all young people and adults across the world should have achieved relevant and recognized proficiency levels in functional and numeracy skills that are equivalent to levels achieved at successful completion of basic education. Obtaining an acceptable level of literacy and numeracy can greatly improve many factors in an individual's life, including improvements in society and community life, education, and career prospects. The ability to read, write, and understand information, can hugely affect the issue of employability.

Education for sustainable development and global citizenship

Education for Sustainable Development and Global Citizenship (ESDGC) is targeted to provide quality education to young people for life in the 21st century. It should not be taken as an additional subject as, rather than a body of knowledge, it is about values and attitudes, understanding, and skills. It is a noble practice that can be implemented throughout schools at a global level, as an attitude to be adapted, a value added system, and a way of life. ESDGC is the bridge between the environment and the inhabitants who live in and from it. It can bring harmony at a global level, and can help to shape rural communities of developing countries for a sustainable healthy livelihood. The ESDGC can help to provide opportunities for both teachers and learners to understand global issues, and to realize how, on a priority basis—personal, local, national, and global—they can critically evaluate and tackle the highly challenging issues of injustice, prejudice, and discrimination.

The ESDGS aims to promote pedagogy and systematic organization of schools to practice the functional reality of education by providing classroom learning and application in the field for the development of rural communities, especially targeting girls and women.

The ESDGS is also targeted to enrich the knowledge of teachers and educators, first to understand new challenges that will be a part of life, such as climatic change and international competition for resources; developing learners' worldview to recognize the complex and international nature of their world; an approach to teaching and learning to which every subject can contribute; and to build the skills that will enable learners to think critically, think laterally, link ideas and concepts, and make informed decisions.

The ESDGS also encourage and promote students' careers and their environment by providing access and opportunity to understand the global scenario of development of developing countries in the field of science and technology, socioeconomic issues, the concept of a universal health system, and implementation of international projects like Sustainable Development Goals to eradicate hunger, poverty, and illiteracy, especially with reference to rural women's development. The three "means of achieving targets" are:

1. build and upgrade inclusive and safe school, and safe school;
2. expand higher education scholarship for developing countries; and
3. increase the supply of qualified teachers in developing countries.

The overall target of SDG4 is to promote access to quality education and other sources of learning opportunity. The most significant aim of SDG4 is to achieve universal literacy and numeracy. Presently, most of the rural areas of developing countries are under the blanket of extreme poverty, insurgency, and social disparity. Children with poor family economic conditions are vulnerable to dropping out of primary education. A wide range of disparity still exists in Western Asia, North Africa, and Sub-Saharan Africa.

Education plays a key role in sustainable development and upgrading the quality of life. In spite of achieving goals of gender equality in primary education, many developing countries still are in the process of struggling in improving the level of education. it is estimated that 57 million children do not attend school, of which more than half live in Sub-Saharan Africa.

With the progress in the SDG4 program, the gender gap in literacy has declined at a steady rate. In Europe and Central Asia, Latin America and the Caribbean, and East Asia and the Pacific, the gender gap has been almost closed. A 17% gender gap exists in South Asia, where the adult male literacy rate is about 79% and adult female literacy rate is about 62%; this is world's largest gender gap. Sub-Sahara Africa and the Middle East and North Africa lag only marginally behind, with a gender gap of 15% points and 14% points, respectively. However, Sub-Saharan Africa has the lowest level of adult female literacy at 57%.

According to the study from the UN Educational, Scientific, and Cultural Organization (UNESCO), COVID-19 has exposed that 584 million youths lack basic literacy skills, and most of them are female. This figure exceeds the 2020 figure of 460 million children lacking basic educational skills [60].

Since the beginning of the COVID-19 pandemic, complete or partial closures have disrupted schooling for an average of 25 weeks, with the highest learning losses in the Latin America and Caribbean region, and in Central and Southern Asia. It has been predicted that the social and economic damage caused by the COVID-19 pandemic could be mitigated by 2024 "if exceptional efforts are made to provide remedial classes and catch-up strategies." According to the UNESCO, UNICEF, the World Bank, and the Organization for Economic Co-operation and Development (OECD) global "Survey on National Education Responses to COVID-19 School Closures," about one in three countries where school have been closed are unable to implement remedial programs post-COVID-19 [61].

Gender equality is a global priority for UNESCO and inextricably linked to promoting the right to education, irrespectively, to boys, girls, adult women, and men. It is closely linked to the right to education for all, and to achieve gender equality in all respect in the society. UNESCO's work on education and gender equality is guided by the UNESCO Strategy for Gender Equality in and Through Education (2019–2025) and the Gender Equality Action Plan (2014–2021, 2019 revision). It is aimed at universal education to benefit all learners equally, and supports targeted action for girls' and women's empowerment across three areas of priority: better data, better policies, and better practices.

2.1.3.3 Neighborhood and environmental health

In creating a sustainable pattern of health, the surrounding environment is a key factor. Environmental health is the branch of public health dealing with the safety and security of a healthy life with reference to the neighborhood (Fig. 2.16).

So, it is necessary to keep the neighborhood clean and fit for quality living. It is the foremost duty of state government to monitor the municipality, healthcare centers, and educational institutes to keep the neighborhood in a harmonious pattern, where an individual, irrespective of cast, gender, and race, should get the opportunity to lead a sustainable pattern of healthy life. The WHO website states:

Environmental health addresses all the physical, chemical, and biological factors external to a person, and all the related factors impacting behaviors. It encompasses the assessment and control of those environmental factors that can potentially affect health. It is targeted towards preventing disease and creating health-supportive environments. This definition excludes behavior not related to environment, as well as behavior related to the social and cultural environment, as well as genetics [62].

According to the WHO environmental health services are defined as:

Those services which implement environmental health policies through monitoring and control activities. They also carry out that role by promoting the improvement of environmental parameters and by encouraging the use of environmentally friendly and healthy technologies and behaviors. They also have a leading role in developing and suggesting new policy areas [63,64].

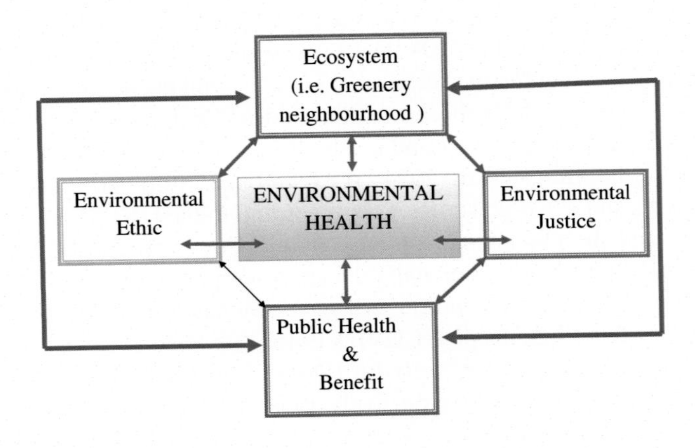

FIG. 2.16

Conceptual interlinking systems of environmental health with various factors.

The WHO Regional Office for Europe considers: "both the direct pathological effects of chemical, social and cultural environ, which includes housing, urban development, land use and transport" [65].

Generally, the intensity of poor health conditions varies with the nature of the neighborhood where we live. Therefore, the social and economic conditions and status of government are closely related to general health status, mortality, birth outcomes, chronic disease, health behaviors, and other risk factors related to ill health [66–69]. Various environmental factors like pollution, air quality (mainly in industrial areas or metro localities), and hazardous chemicals effect both mental and physical health.

Certain bad social practices, such as smoking or alcohol addiction, can have considerable impacts on heath. Such problems in an individual can be motivated by living in a neighborhood that lacks safe areas for exercise, and surroundings where intensive tobacco and alcohol advertising targets poorer and minority youths; protection from which can bring self-realization to keep them free from unsocial habits and lead a healthy life.

How to create healthier neighborhoods

Public health and neighborhoods are complementary with each other. Surroundings can have a have direct impact on individual as well as community health. In particular, children, during their developing stage, need a favorable and clean environment both at home and in their neighborhood. Generally, children are vulnerable to unhealthy conditions in neighborhoods. Below are some of the important factors to be taken care of by both public and private sectors, while undertaking any social housing project:

(i) Developing affordable food markets near social housing is one of the better options to save time and money, which can be used for developing a healthy livelihood. Poverty and health are complementary to each other. Therefore, poverty eradication is the one of the major issues of government, international agencies, and civil society. The most effective way to achieve this is to focus on enabling institutional environments, rather than direct initiatives to reduce poverty. In this connection, development of a neighborhood market is one of the better factors related to developing social housing (Fig. 2.17).

(ii) Community revitalization promotes neighborhood economic development and improves individual and community health. In addition, it also helpful in developing in children the inspiration to lead a healthy life. The revitalization of community can only be possible by developing primary health centers, nurseries, and primary schools, development of parks to refresh children's mental health conditions, and psychological and pedagogical support.

Revitalization of rural communities is also a burning issue for developing countries. Rural communities and spaces act as the interface between human societies and the natural world. The rural community has vital functions in harvesting and promoting the natural resources that are linked to the development of communities of all sizes and urbanity. So, it is necessary to develop primary health centers, nurseries

FIG. 2.17

Metro mall in a housing complex, Delhi, India.

and primary schools, children's recreation centers neighboring rural communities. Rural revitalization requires a transformative approach that considers all aspects of making rural areas good places to live and work for present and future generations.

(i) Community organization is a process of bringing both rural and urban community people together to work collectively to improve neighborhoods. In this process, the community identifies needs, and accordingly works for development on a priority basis. Community organization should occur in geographically, psychosocially, culturally, spiritually, and digitally bounded communities. Community organization can be promoted by implementing community development projects, developing nonconventional energy sources, especially at village level, a community transport system within a neighborhood to promote people's activities, and rural community empowerment with especial reference to rural women and their respective livelihoods.

(ii) In general, urban communities face the problem of environmental pollution leading to damage to the physical environment, including water resources, air quality, and neighborhood ecosystems. This can reduce the quality of life on Earth. The environmental impact of nonrenewable resources, higher levels of pollution, global warming, and potential loss of environmental habitats not only effect the community but also bring health problems at an individual level.

The following are the best and simplest ways to avoid community and neighborhood environmental pollution:

(a) Recycle household waste by creating organic compost, biogas, or conservation of rural resources and landfill space.

(b) Establishing a volunteer service to keep the community and neighborhoods clean.
(c) Reduce or stop using nondegradable plastic or polythene carriers by replacing with environmentally degradable carrier bags.
(d) Educating the people of urban and rural communities on environmental pollution and preventive measures.
(e) Conserve water and avoid runoff and wastewater that eventually end up in the ocean or rivers. Promote rainwater harvesting practices in communities and neighborhoods.
(f) Promote the process technology for using seafood as a supplement in dietary habits.
(g) Use of cycle-like devices for transport processes within communities and around neighborhoods.
(h) Promote tree planting within neighborhoods to get relatively clean air.
(i) Stop toxic chemical discharge into the environment.
(j) Encourage and promote long-lasting light bulbs, preferably LD technology-based light systems.
(k) Strategies to minimize residential segregation, such as through zoning, and affordable cluster housing complexes with quality school and employment opportunities, and enforcement of fair housing laws.

References

[1] FAO, IFAD, UNICEF, WFP and WHO. The state of food security and nutrition in the world (SOFI). In: Transforming food systems for affordable healthy diets. Rome, Italy: FAO; 2020. p. 320. https://doi.org/10.4060/ca9692en.

[2] International Labour Organisation (ILO). Global employment trends for women; 2009. Available from: http://www.ilo.org/wcmsp5/groups/public/—dgreports/—dcomm/documents/publication/wcms_103456.pdf.

[3] Poverty defined according to the United Nations. http://www.un.org/sustainabledevelopment/poverty/.

[4] The definition of extreme poverty has just changed—here's what you need to know. https://www.odi.org/comment/9934-definition-extreme-poverty-has-just-changed-here-s-what-you-need-know.

[5] World Bank Group (2020) Poverty and shared prosperity. https:www.worldbank.org>publication>poverty-and.

[6] Anon. Poverty and inequalities in Kenya/SID East Africa regional office. Development 2007;50:182–3. https://doi.org/10.1057/palgrave.development.1100377.

[7] Anon. Sixty percent population living below poverty line: World Bank report. Business Recorder; 2013.

[8] UNIFEM (now UN Women). Progress of world's women. Who answers to women? Gender and accountability; 2009. p. 36.

[9] Anon. Female genital mutilation/cutting: a global concern. New York: UNICEF; 2016.

[10] Nestle M. Food lobbies, the food pyramid, and U.S. nutrition policy (PDF). Int J Health Serv 1993;23(3):483–95.

[11] U.S. Department of Health and Human Services, Office of Disease Prevention and Health Promotion. What is healthy people?.
[12] International Labour Organisation. Working in the rural areas in the 21st century: reality and prospects of rural employment in Latin America and the Caribbean, regional Office for Latin America and the Caribbean (Geneva, Thematic Labour Overview No 3, 2017); 2017. p. 38.
[13] Anon. International Labour Office: Decent work for food security and resilient rural livelihoods, Portfolio of policy guidance notes on the promotion of decent work in the rural economy. Geneva: Sectoral Policies Department; 2016.
[14] Anon. International Labour Office: World employment social outlook 2016: transforming jobs to end poverty (Geneva); 2016.
[15] Anon. The average rate of female agricultural land ownership is less than 20 per cent in developing countries. Rome: FAO; 2010.
[16] Food and Agriculture Organization. The state of food and agriculture 2010–2011—I: women in agriculture: closing the gender gap for development. Rome: FAO; 2011.
[17] Food and Agriculture Organization, IFAD and ILO. Gender dimensions of agricultural and rural employment: differentiated pathways out of poverty—status, trends and gaps. Rome: FAO; 2010.
[18] Gopalakrishnan R, Sukthankar A. Freedom of association for women rural workers: a manual. Geneva: ILO; 2012.
[19] International Labour Organization. Meeting of experts on violence against women and men in the world of work. Geneva: ILO; 2017.
[20] International Labour Organization. Walk free foundation and IOM: global estimates of modern slavery. Geneva: ILO; 2017.
[21] Adepoju A. Migration in West Africa. In: A paper prepared for the policy analysis and research programme of the global commission on international migration. Global Commission on International Migration; 2005. http://www.gcim.org/attachements/RS8.pdf. [Accessed 20 August 2014].
[22] Beauchemin C, Bocquier P. Migration and urbanisation in francophone Beauchemin West Africa: an overview of the recent empirical evidence. Urban Stud 2004;41(11):2245–72.
[23] Collinson M, Tollman S, Kahn K, Clark S, Garenne M. Highly prevalent circular migration: households, mobility and economic status in rural South Africa. In: Tienda M, Findley S, Tollman S, Preston-Whyte E, editors. Africa on the move: African migration and urbanization in comparative perspective. Johannesburg: Wits University Press; 2006. p. 194–216.
[24] Reed HE, Andrzejewski CS, White MJ. Men's and women's migration in coastal Ghana: an event history analysis. Demogr Res 2010;22:771–812.
[25] Sudarkasa N. Women and migration in contemporary West Africa. In: The Wellesley Editorial Committee, editor. Women and National Development: The complexities of change. Chicago, IL: University of Chicago Press; 1977. p. 178–89.
[26] Hill P. Rural Hause: a village and a setting. Cambridge: University Press; 1972.
[27] Singh KN. Urban development in India. Abhinav Publications; 1978, 1 January, ISBN:978-81-7017-080-8.
[28] Anon. Urbanization in India faster than rest of the world. Hindustan Times; 2007, 27 June. Archived from the original on 26 November 2011.
[29] Business Standard. Victims of urbanization: India, Indonesia and China., 2012, 15 June, Rediff.com.

[30] Sivaramakrishnan KC, Dasgupta B, Buch MN. Urbanization in India: basic services and People's participation. Concept Publishing Company; 1993, 1 January. p. 2, ISBN:978-81-7022-480-8.
[31] Gulati L. Women, work and migration in Asia. In: Agarwal A, editor. Women and migration in Asia. New Delhi: Sage Publications; 2006.
[32] Arya S, Roy A. Poverty, gender and migration. New Delhi: Thousand Oaks: London: Sage Publications; 2006.
[33] Premi MK. Patterns of internal migration of females in India. In: Center for the Study of Regional Development: Occassional Paper No 15. New Delhi: Jawaharlal Nehru University; 1979.
[34] Shanti K. Issues relating to economic migration of females. Indian J Labour Econ 1991;34(4):335–46.
[35] Hugo GJ. Migrant women in developing countries. In: Internal migration of women in developing countries. Proceedings of the United Nations; 1993.
[36] Gracia B. Women, poverty and demographic change. New York: Oxford University Press; 2000.
[37] Wille C, Passl B, editors. Female labour migration in South-East Asia: change and continuity. Bangkok: Asian Research Centre for Migration; 2001.
[38] Thanh-Dan T. Gender, international migration and social reproduction: implications for theory, policy, research and networking. Asian Pac Migr J 1996;5(1):27–42.
[39] Yeoh B, Huang S, Gonzales J. Migrant female domestic helpers: debating the economic, social and political Impacts in Singapore. Int Migr Rev 1999;33(1):114–36.
[40] Mahapatro SR. Patterns and determinants of female migration in India: insights from census. Working paper 246. Bangalore: Institute for Social and Economic Change (ISEC); 2010.
[41] Asis MMB. Asian women migrants: going the distance, but not far enough. Migration Information Source (the Online Journal of Migration Policy Institute); 2003, 1 March.
[42] Lingam L. Locating women in migration studies: an overview. Indian J Soc Work 1998;59(2):715–27.
[43] Conell J. Status on subjugal women migration and development in the South Pacific. Int Migr Rev 1984;18(4):966.
[44] Kasturi L. Poverty, migration and women's status. In: Majumdar V, editor. Women workers in India: studies in employment and status. New Delhi: Chanakya Publication; 1990.
[45] Neetha N. Making of female breadwinners: migration and social networking of women domestics in Delhi. Econ Polit Wkly 2004;24(April):24–30.
[46] Saradamoni K. Crisis in fishing industry and women's migration: the case of Kerela. In: Schenk-Sandbergen L, editor. Women and seasonal labour migration. Delhi: Sage Publications; 1985.
[47] Institute of Medicine (US) Committee on Monitoring Access to Personal Health Care Services, Millman M. Access to health care in America. The National Academies Press, US National Academies of Science, Engineering and Medicine; 1993, ISBN:978-0-309-04742-5. https://doi.org/10.17226/2009. 25144064.
[48] Andersen RM. Revisiting the behavioral model and access to medical care: does it matter? J Health Soc Behav 1995;36(1):1–10.
[49] Chapman JL, et al. Systematic review of recent innovations in service provision to improve access to primary care. Br J Gen Pract 2004;54(502):374–81.

[50] Donabedian A. Models for organizing the delivery of personal health care services and criteria for evaluating them. Milbank Mem Fund Q 1972;50(4, Pt 2):103–54.
[51] Gulliford M, Figueroa-Munoz J, Morgan M, et al. What does 'access to health care' mean? J Health Serv Res Policy 2002;7(3):186–8.
[52] Parker AW. The dimension of primary care: blueprint for change. In: Andreopoulos S, editor. Primary care: where medicine fails. New York: Wiley; 1974. p. 15–77.
[53] Penchansky R, Thomas JW. The concept of access: definition and relationship to consumer satisfaction. Med Care 1981;19(2):127–40.
[54] Halseth G, Ryser L. Trends in service delivery: examples from rural and small town Canada, 1998 to 2005. J Rural Community Dev 2006;1:69–90.
[55] The Collective. Chinese healthcare: the rural reality., 2018, https://www.coresponsibility.com.
[56] Kaunain Sheriff M. Government report: shortfall of 76.1% specialist doctors at CHCs in rural areas. The Indian Express; 2021. August 16.
[57] Thayyil J, Jeeja MC. Issues of creating a new cadre of doctors for rural India. Int J Med Public Health 2013;3(1):8. https://doi.org/10.4103/2230-8598.109305.
[58] Duggal R. Healthcare in India: changing the financing strategy. Soc Policy Adm 2007;41(4):386–94. https://doi.org/10.1111/j.1467-9515.2007.00560.
[59] Bhardwaj G, Monga A, Shende K, Kasat S, Rawat S. Healthcare at the bottom of the pyramid an assessment of mass health insurance schemes in India. J Insur Inst India 2014, April 1;1(4):10–22.
[60] United Nations. 100 million more children fail basic reading skills because of COVID-19., 2021, https://news.un.org/en/story/2021/03/1088392.
[61] UNICEF. UNESCO, UNICEF, World Bank and OECD report documents education responses to COVID-19 in 142 countries. http://www.unicef.org/press-releases/1-3 countries.
[62] World Health Organisation. Environmental health., 2015, https:www.who.int/Health topics.
[63] Brooks Bryan W, Gerding Justin A, Landeen E, et al. Environmental health practice challenges and research needs for U.S. health departments. Environ Health Perspect 2019;127:125001.
[64] MacArthur I, Bonnefoy X. Environmental health services in Europe 1. An overview of practice in the 1990s. WHO Reg Publ Eur Ser 1997;76(vii–xii):1–177.
[65] Novice R, editor. Overview of the environment and health in Europe in the 1990s. World Health Organization; 1999, March 29. PDF.
[66] Sampson R, Morenoff J, Gannon-Rowley T. Assessing "neighborhood effects": social processes and new directions in research. Annu Rev Sociol 2002;28:443–78.
[67] Yen I, Syme SL. The social environment and health: a discussion of the epidemiologic literature. Annu Rev Public Health 1999;20:287–308.
[68] Pickett KE, Pearl M. Multilevel analyses of neighbourhood socioeconomic context and health outcomes: a critical review. J Epidemiol Community Health 2001;55(2):111–22.
[69] Robert SA. Socioeconomic position and health: the independent contribution of community socioeconomic context. Annu Rev Sociol 1999;25:489–516.

CHAPTER

The life cycle vulnerabilities of rural women 3

The biological life cycle is a common process for all men and women. The continuum of an individual's life can be divided into several stages, each characterized by certain features. However, a woman's life cycle is different from a man's from a reproduction point of view. The life cycle of a woman is divided into infancy, reproductive stage, climacteric period, and elderly years. Gender inequalities tremendously harm the physical and mental health of millions of girl and women across the globe. In particular, women residing in rural areas face problems due to poverty, food insecurity, hunger, social exclusion, lack of education, availability of only extremely poor primary health services, and lack of transport facilities to access nearby health centers.

3.1 Women's rights

"The enjoyment of the highest attainable standard of health is one of the fundamental rights of every human being without distinction of race, religion, political belief, economic or social condition." Everyone has the right to privacy and to be treated with respect and dignity [1].

It is the legal obligation of state governments to provide access for each and every citizen to timely health services for their entire life cycle, that is, acceptable and affordable healthcare of desirable quality, as well as to ensure determinants of health, including safe drinking water useful for all household purposes, sanitation, nutrient-rich food, digital healthcare services, education, and gender equality.

It should be an obligation and duty of states to provide "adequate resources" to keep attention on the timely fulfillment of the goal by monitoring through various international human rights mechanisms, such as the Universal Periodic Review, or the Committee on Economic, Social, and Cultural Rights. As per 29 December 2017, WHO statement: "A right-based approach to health requires that health policy and programmes must prioritize the needs of those furthest behind first towards greater equity, a principle that has been echoed in the recently adopted in the 2030 Agenda for Sustainable Development and Universal Health Coverage" (SDGs) [2].

3.2 Why is rural women's health the first priority?

Rural women constitute one-fourth of the world's population. They are the vital force for rural development thru the well-being of families, communities, and

Healthcare Strategies and Planning for Social Inclusion and Development. https://doi.org/10.1016/B978-0-323-90447-6.00002-3

economies, and achievement of the Sustainable Development Goals (SDGs). Therefore, it is the foremost duty of government, politicians, and policy makers to focus on to take care of and overcome the life cycle vulnerabilities of rural women. Generally, the life cycle of a woman is categorized into infancy, puberty, reproductive age, climacteric period, and elderly years; in addition, pregnancy and delivery are generally included as life events unique to women. Urban women have access to healthcare services during their different life cycle stages, but women residing in rural areas and remote localities have poor access to health services during their lifetime (Fig. 3.1).

One of the most chronic features of rural health problems is the random development and distribution of rural villages in small clusters, isolated from each other, with significant distance and poor access to urban life due to lack of affordable transport systems. In addition, the shortage of healthcare professionals is a global issue and has been a challenging problem in monitoring health of rural people, especially of girls and women. Mostly, in a rural community, the number of elderly people and children, and unemployed youth is more, and they need healthcare services. But lack of proper healthcare infrastructure, poor access to communications, and lack of primary and secondary education make rural areas unattractive to healthcare workers, resulting in an overwhelming disparity in doctor/patient ratio in urban areas compared to rural areas. Consequently, retention of healthcare workers in rural or remote areas is a vital task of state governments, particularly of developing countries. In addition, other factors like health worker migration, professional/specialty inequalities, institutional inequities, humanitarian crises like war and civil repression, and outbreak of contagious communicable disease like

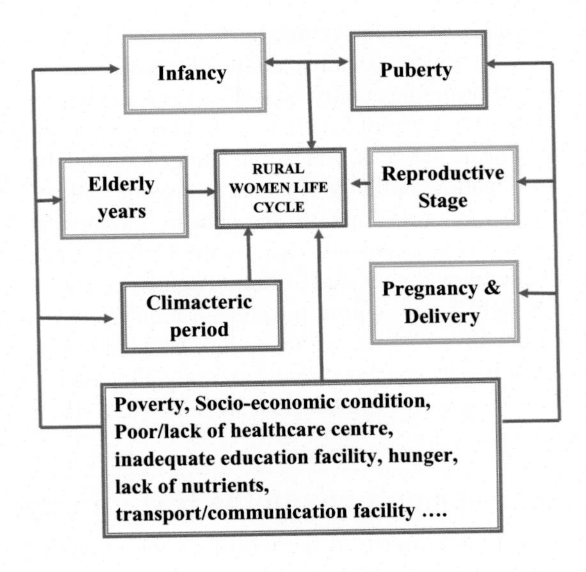

FIG. 3.1

Rural women's lifecycle facing various problems.

Table 3.1 Rural population in different countries rank wise.

No.	Name of the country	Population	Year
1	India	892,321,700	2018
2	China	568,902,300	2018
3	Pakistan	134,404,300	2018
4	Indonesia	119,578,600	2018
5	Bangladesh	102,248,100	2018
6	Nigeria	97,263,560	2018
7	Ethiopia	86,546,260	2018
8	Vietnam	61,223,240	2018
9	United States	58,052,590	2018
10	Philippine	56,624,700	2018

(Source: World Bank staff estimates based on the United Nations Population Division's World Urbanization prospects: 2018 Revision.)

COVID-19 and HIV/AIDS catalyze the poor healthcare conditions in rural/remote areas. "Rural population" is a variable term from country to country and is defined by the country's statistical office (Table 3.1).

3.3 World conference on women

In order to overcome women problems, the United Nations organized a series of world conferences on women; these took place in Mexico City in 1975, Copenhagen in 1980, Nairobi in 1985, and Beijing in 1995. The fourth World Conference reviewed the past agendas and placed stress on gender equality on a priority basis. Among the various issues, resolving women's health-related problems during the life cycle is also a priority on the agenda, especially with reference to rural women's health.

Following are a few important issues on which special emphasis is given to women's health with special reference to rural women's vulnerability to ill health during the different phases of life.

3.3.1 Women's rights

Women have a right to the enjoyment of the highest standard of physical and mental health, which must be guaranteed throughout her lifetime, equal to that of men. There are many health problems common to both men and women, but women may experience them differently, due to both genetic differences and social exclusion. Sound health is not only the absence of disease, but is also a state of complete physical, mental, and social well-being. Women's health is closely linked with their emotional, social, and physical well-being, and is dependent on social inclusion, and the political and economic context of their life, as well as their biology. The major barriers to achieving the standard of health equality are gender inequality, geographical region, and race and ethnicity. Good health is essential to continue a productive life in a dignified manner, and the right of all women to take care of all aspects of their health, in particular their own fertility, is fundamental to their freedom and empowerment.

3.3.2 Ethnic minorities

Even at international level, there is no common definition of "ethnic minorities." However, an ethnic group commonly shares a common sense of identity and common characteristics such as language, religion, tribe, nationality, race, or a combination thereof. The term "ethnic minority" commonly refers to an ethnic or racial group within a given country in which they occupy a nondominant position [9].

Although, women are not under ethnic group but still innumerable challenges and issues that women face that concern physical and mental health. The problems like lack of education, improper health facilities, gender discrimination, gender pay gap, unequal right, sexual harassment, dowry-related problems, and domestic violence are still persisting, mostly in rural area of developing country. So, in national and international forums, women have emphasized that to attain optimal health throughout their life cycle, equality, including the sharing of family responsibilities, development, and peace are necessary conditions.

3.3.3 Access to women's health

Mostly, in rural area women take the overall burden of family, including taking care of health of elderly members and children. So, it is necessary to provide them basic access to comprehensive and quality health care services. The health of women and girls is of particular concern because, in many societies, they are disadvantaged by discrimination rooted in sociocultural factors. For example, women and girls face increased vulnerability to HIV/AIDS. So, women need basic quality health resources, which include primary health services for prevention and treatment of childhood diseases, malnutrition, anemia, and diarrheal diseases. Women face discrimination in protection, promotion, and maintenance of their health. In particular, in many developing countries, inadequate emergency obstetric services are of concern. Most maternal death resulting from direct obstetric complications are avoidable [3], and even complicated obstetric cases can be treated with the financial support of eight interventions identified by the World Health Organization (WHO), the United Nations Children's Fund (UNICEF), and the United Nations Population Fund (UNFPA) that, taken together, are known as emergency obstetric care (EmOC) [4–6].

In many developing and least developed countries, health policies and programs often perpetuate gender stereotypes, fail to deal with socioeconomic disparities and other differences among women, and may not fully take account of the lack of autonomy of women regarding their health. Furthermore, gender bias attitudes in health systems are a chronic problem, especially for women residing in rural areas.

3.3.4 Budgetary provision for women's health

In many developing countries and least developed countries, the limited budget for public health results in deteriorating systems. In addition, privatization of healthcare system without appropriate guarantees of universal access to affordable healthcare further reduces healthcare availability. Both in developed countries and developing countries, for-profit healthcare is more expensive and often lower quality than not-for-profit or government care, with much higher overhead costs. Partial privatization draws off resources from the public system, increases costs overall, and introduces inequalities into the healthcare system. Privatization of health services adversely affects rural people, especially rural women, as it results in fiscal devolution: the rural primary health service is transformed into a fee-for-service system, dependent on the availability of local resources. This situation not only directly affects the health of girls and women, but also results in placing disproportionate responsibilities on women and their key role in family care and community development.

3.3.5 Safety and security for women's health

Women residing in rural localities, especially of developing countries and least developed countries, face threats to their safety and well-being. Rural women continue to face systematic and persistent barriers to the full enjoyment of their human rights and in fact, in many cases, conditions have deteriorated. In many states, the right to health of rural women is ignored while planning for states' budget and investment strategies at all levels. Even where laws and policies consider rural women's situation and foresee special measures to address them, they are often not implemented. Most rural women are severely in the grip of poverty, social exclusion, family and social violence, economic dependence, negative attitudes toward women and girls, and race and other form of discrimination, which prevents them from exercising their rights in a timely fashion. The lives and livelihoods of women are constantly threatened and made unsafe through lack of food, and inequitable distribution of food, for girls and women in the household, inadequate access to safe water and sanitation facilities, and deficient housing conditions.

3.3.6 Provision of sanitation

Proper sanitation facilities (for example, toilets and latrines) promote health because they allow people to dispose of their waste appropriately. In most developing countries and least developed countries, many people do not have access to sanitation facilities, resulting in improper waste disposal. Good health is essential to leading a productive and fulfilling life, and the right of all women to control all aspects of their own health.

3.3.7 Preference for males in the family

Son preference is an abnormal attitude in many developing countries like India. In this case, if a female baby is born, she is subject to poor access to quality nutrition and healthcare services, which may endanger her current and future health and well-being.

Early marriage, pregnancy, and child-bearing carry serious health risks for mother and child. This practice, which may lead to female genital mutilation and pregnancy-related health problems, can have emotional and social consequences and pose a financial burden to the household. Generally, at an early age, females tend to have no access to financial resources and suffer restricted mobility; they are, therefore, less likely to leave home to socialize with others, and have limited access to information on reproductive health, contraception, HIV, and other sexually transmitted infections (STIs). This power differential can also limit girls' ability to negotiate contraceptive or condom use, putting them at high risk for contracting STIs and HIV.

3.3.8 Reproductive health

Reproductive health, or sexual health, addresses the reproductive processes, functions, and system at all stages of life. As per the UN agencies report, both sexual and reproductive health address physical as well as psychological well-being vis-à-vis sexuality [7].

3.3.9 Family structure and rights of couples

It is the right of the couple to plan their family on the basis of personal financial status and family conditions. On the basis of mutual understanding, the couple has the right to determine the age gap of their respective forthcoming generation. Generally, the cultural heritage of family and parental care often guide married children how to plan the family for developing a happy home. Elderly family members take care of married children's life cycle and also socially and financially help the newly married children for establishing a well-planned new family. The elderly family members also guide their respective children how to avail themselves of various opportunities and advice from family planning healthcare professions for a comfortable and successful family life with good health.

However, due to lack of education, poverty, and distance, rural women face limited reproductive health services. The resulting lack of care can challenge rural women's reproductive autonomy. Their reproductive choices may also be limited by the added impact of rural values, norms, and belief systems regarding sexual health and the patient-physician relationship, due to lack of education, poor job opportunities, low income, more children, and greater family caretaking responsibility than their urban counterparts. Rural women are more likely to marry at a younger age. The problem of poverty, low population density, and lack of childcare and other services in many rural areas reinforces traditional roles for women. Rural women have less accesses to preventive care than women in urban areas and have the highest rates of chronic disease.

3.3.10 Pregnancy complications

Complications related to pregnancy and childbirth is among the leading causes of mortality and morbidity of women of reproductive age in many developing countries. A similar problem exists to a certain degree in some countries with economic transition. Unsafe abortions threaten the lives of a large number of women, representing a grave public health problem, as it is primarily the poorest and youngest who take the highest risk. Most of these deaths, health problems, and injuries are preventable through improved access to adequate healthcare services.

3.3.11 Pregnancy-related problems

Biological changes caused by HIV, including systemic illness, stress, and weight loss, may affect the function of reproductive organs and result in infertility. Newly diagnosed HIV infection may cause psychological trauma and decrease in sexual drive and sexuality. Women, who represent half of all adults newly infected with HIV/AIDS and other sexually transmitted diseases, have emphasized that social vulnerability and the unequal power relationships between women and men are obstacles to safe sex, in their effort efforts to control the spread of sexually transmitted diseases.

3.3.12 Gender-based violence

Gender-based violence threatens women's health worldwide, adding to the global burden of diseases [8–10].

Generally, women are victims of poverty, powerlessness, and overload with domestic and other related work, which often lead to domestic misunderstanding and violence as well as substance abuse, among other health issues of growing concern to women. These sorts of unwanted incidents are commonly noticed in rural areas, leading to adverse effects on women's lives and their children's. Occupational health issues are also growing in importance, as a large number of women work in low-paid jobs in either the formal or the informal labor market under tedious and unhealthy conditions, and the number is rising. Frequently, these women workers suffer from cancers of the breast and cervix and other cancers of the reproductive system, as well as infertility effecting growing numbers of women, all of which may be preventable and curable, if detected early.

The World Health Organization has a gender-based approach to public health to resolve the various issues related to gender-based violence and health vulnerability and risks of rural women and girls that are identified and addressed in health service delivery. In addition, many other international agencies are involved in attempting to resolve and improve the overall well-being of rural women through advocating for social protection, decent work, and elimination of difference in daily wages for men and women of rural communities. The United Nations Fund for Population Activities (UNFPA) does in-depth work on reproductive rights and sexual and reproductive health to cover the targets for Sustainable Development Goals (SDGs) (Table 3.2).

3.3.13 Infertility

Infertility affects a growing numbers of women, but may be preventable, or curable, if detected early. Currently, infertility is the most important issue following cancer and cardiovascular diseases. Although it is not a danger to life, it can bring chronic

Table 3.2 Various aspects of SDG3 related to women's health and prevention.

SDG 3.2	By 2030, end preventable deaths of newborns and children under 5 years of age, with all countries aiming to reduce neonatal mortality to at least as low as 12 per 1000 live births and under-five mortality to at least as low as 25 per 1000 live births
SDG 3.3	By 2030, end the epidemics of AIDS, tuberculosis, malaria, and neglected tropical diseases, and combat hepatitis, water-borne diseases, and other communicable diseases
SDG 3.7	By 2030, ensure universal access to sexual and reproductive healthcare services, including for family planning, information, and education, and the integration of reproductive health into national strategies and programs
SDG 3.8	Achieve universal health coverage, including financial risk protection, access to quality essential healthcare services, and access to safe, effective, quality, and affordable essential medicines and vaccines for all
SDG 5.2	Eliminate all forms of violence against all women and girls in the public and private spheres, including trafficking and sexual and other types of exploitation
SDG 5.3	Eliminate all harmful practices, such as child, early, and forced marriage and female genital mutilation
SDG 5.6	Ensure universal access to sexual and reproductive health and reproductive rights as agreed in accordance with the Programme of Action of the International Conference on Population and Development and the Beijing Platform for Action and the outcome documents of their review conferences

unhappiness in the life cycle of a woman. Due to the serious impact of infertility on human reproductive health of both men and women, the issue has attracted much concern [11,12]. Some of the main reasons for the increase in infertility are various environmental stresses and the current highly modernized life style. A decade back, infertility in couples worldwide accounted for 10%–15% of all married couples [13].

3.3.14 Population aging

Globally, human longevity is increasing. By 2050, the world's population aged 60 years and older is expected to total 2 billion, up from 900 million in 2015 [14]. In 2018, 125 million people were aged 80 years or older. It is predicted that by 2050, there will be almost 120 million in China and 434 million people in this age group globally. By 2050, 80% of all older people will live in low- and middle-income countries [14].

The natural phenomenon of shifting in distribution of a country's population toward older ages is known as "population aging." In Japan, 30% of the population are already over 60 years old. It is expected that low- and middle-income countries will face population aging. By the middle of the century, many countries, for example Chile, China, the Islamic Republic of Iran, and the Russian Federation, will have a similar proportion of older people to Japan.

Population aging may cause huge problems due to lack of transport options, social isolation, and loneliness for elderly people. Generally, the longevity of women is comparatively greater than that of men. Therefore, elderly rural women will be severely affected by the problem of population aging: the survival rate of older women will increase, and they will need special attention for healthcare. The long-term health prospects of women are influenced by changes at menopause, which, in combination with life-long conditions and other factors, such as poor nutrition and lack of physical activity, may increase the risk of cardiovascular disease and osteoporosis. Other diseases of aging and the interrelationships of aging and disability among women also need particular attention.

3.3.15 Environmental hazards and rural women's health

The rural areas of developing countries and least developed countries are also increasingly exposed to environmental health hazards owing to environmental catastrophes and degradation. Women are prone to environmental hazards, contaminants, and climatic changes in different ways to men. This may be due to the higher proportion of body fat (>10%) of women as compared to men. When the body is exposed to a toxic environment, women retain more toxic substances as compared to men, which results in health vulnerability and frequent illness. Hormonal differences may also affect the way a person's body responds to chemicals. Rural air pollution sources are associated with greater cancer mortality. For example, rural coal mining areas have higher total, cancer, and respiratory disease mortality rates. Rural women engaged in agricultural work are regularly exposed to highly toxic pesticides. Reproductive effects that have been associated with pesticide exposure in women are decreased fertility, spontaneous abortions, stillbirth, premature birth, low birth weight, developmental abnormalities, ovarian disorders, and disruption of the hormonal function [15–17].

3.4 Life cycle vulnerabilities of rural women

Commonly, a woman passes though different stages before completion of elderly stages. However, the pattern for women from urban areas and rural area is entirely different. Although, rural women play a significant role in achieving a successful livelihood, in spite of poor food security, low income, adverse climatic conditions, and less empowerment as a family leader—and additionally, rural women act catalytically in achievement of transformational economic, environmental, and social changes required for sustainable development—nevertheless, limited access to credit, healthcare, and education are among the many challenges they face with other family members. The following are the different stages of a rural woman before the completion of life.

3.4.1 Infancy stage

"Infant" is derived from the Latin word *infans*, meaning "unable to speak" or "speechless." Commonly, "baby" is used as a synonym in place of infant. An infant refers to a newborn, who is only hours, days, or up to 1 month old (before birth, the term "fetus" is used). The various stages an infant passes through are known as the "infancy stage." This is an extremely critical phase of intense care. However, rural women under the constraints of poverty are not in a position to nurture the infant to grow up as a healthy child.

Women in rural communities face higher rates of life-threatening complications during or after childbirth than mothers in urban areas. Due to nonavailability of quality primary health centers, lack of emergency facilities, and poor transport services, the delivery period is frightening and dangerous, and may result in the death of mother and/or infant.

Infant mortality is the death of children under the age of 1 year. Generally, infant mortality rate is expressed as infant deaths per 1000 live births. Overall, infant mortality rates have significantly decreased all over the world (Fig. 3.2).

In developed countries, such as the United States, infant mortality is mostly caused by congenital disabilities, preterm birth and low birth weight, maternal pregnancy complications, sudden infant death syndrome (SIDS), and injuries (such as suffocation). The major causes of global infant mortality are neonatal encephalopathy (problems with brain function due to lack of oxygen during birth), infections, complications of preterm birth, lower respiratory infections, and diarrheal diseases. The mortality rate in the United States is 5.8, which is remarkably

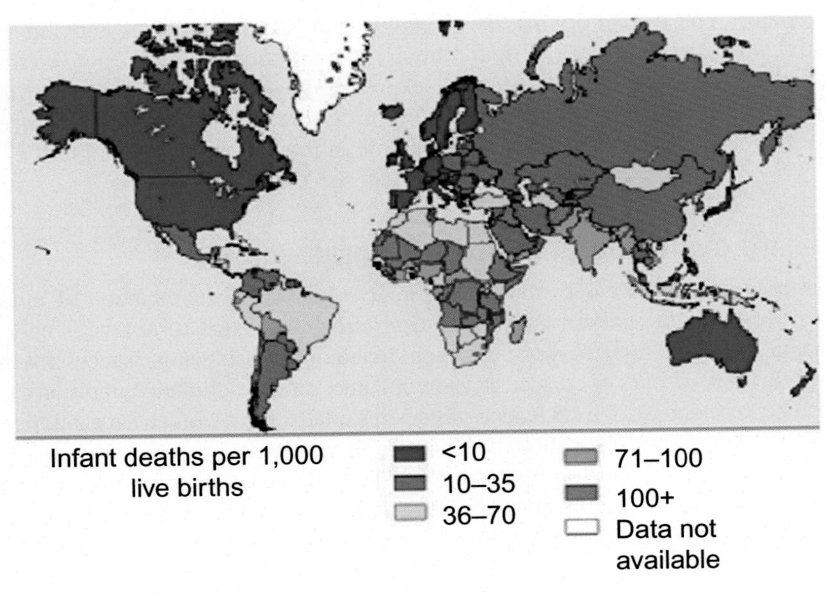

FIG. 3.2

Infant mortality rate May 19, 2021 (infant mortality rate per 1000 live births). Infant mortality rate is the probability of a child born in a specific year or period dying before reaching the age of 1 year.

higher than other developed countries. Various factors like poverty, malaria, malnutrition, undeveloped infrastructure, and poor health facilities also cause high mortality rates.

Globally, Afghanistan has the highest infant mortality rate, at 110.6. This is mainly due to the unevenly and widely spread communities. In addition, most people travel by foot, and accessing healthcare is very difficult, especially for pregnant women and young babies. There is also a significant lack of health education in Afghanistan, so pregnancy complications are often ignored. Afghanistan has a fertility rate of 4.412 births per woman, well above the global average of 2.4, and birth rate of 37.9, the 12th highest globally.

Monaco has the lowest infant mortality rate, at 1.8. This is mainly due to the extreme wealth of the small country, highly educated mothers, and a sound healthcare system.

Infant mortality rates are low in most EU countries, with an average of less than 3.5 deaths per 1000 live birth across EU countries in 2018 [5]. However, a small group of countries (Malta, Romania, Bulgaria, and the Slovak Republic) still have infant mortality rates of 5 deaths per 1000 live births. Over the past few decades, all European countries have made remarkable progress in reducing infant mortality [18].

3.4.2 Infant mortality

3.4.2.1 Types of infant mortality

On the basis of the health condition of mother and infant, the immediate delivery process and healthcare services during the first 1 month period can be categorized under the following headings.

3.4.2.2 Perinatal mortality

Perinatal mortality (PNM) is defined as the late death of a fetus (22 weeks gestation to birth) or neonate during the advanced stage of pregnancy. As per the statement of the World Health Organization, perinatal mortality is the "number of stillbirths and deaths in the first week of life per 1000 total births, the perinatal period commences at 22 completed weeks (154 days) of gestation, and ends seven completed days after birth" [19]. Other definitions proposed by different healthcare agencies are also applicable [20].

In the UK, the infant mortality rate is about 8 per 1000 live births. This figure varies significantly by social class, with the highest rates noticed in Asian women. In 2013, the infant mortality number at a global level was about 2.6 million neonates in the first month of age down from 4.5 million in 1990 [21].

In 2018, as reported by the WHO, there was a significant decline in annual infant deaths, amounting to 4.0 million with reference to 8.7 million in 1990 [22].

3.4.2.3 Neonatal mortality

Neonatal mortality is newborn death, within 28 days postpartum. This occurs mainly due to lack of basic medical care during pregnancy or after delivery. About 40%–60% infant mortality has been reported in most of the developing countries [23].

The World Health Organization report in 2019 declared that, in the first month of infancy, globally, 2.4 million children died. During this period, the newborn cases were about 6700 per day, amounting to 47% of all child deaths under the age of 5 years [24].

With the implementation and progress of Sustainable Development Goals (SDGs) the neonatal deaths in the age group of below 5 years had been remarkably reduced to 2.4 million in 2019 with reference to neonatal deaths cases of 5.0 million in the year 1990. The reduction in neonatal mortality during the period 1990 to 2019 has been noticeably slower as compared to postneonate under-five age group mortality. In his connection, the neonatal deaths in the under-five group decreased to 36%, which was remarkable as compared to past data.

Meanwhile, SDG3, with its 13 subtargets for universal healthcare, has resulted in a significant reduction in neonatal deaths worldwide. In 2015, the neonatal death rate under age 5 years worldwide was 43 deaths per 1000 live births. About 43% reduction in child death under 5 years was noticed over the value of neonatal deaths during 2000. But despite progress in every region, still, wide disparities persist. Sub-Saharan Africa still faces substantial rates of under-five mortality, with a value of 84 deaths per 1000 live births in 2015. This value is about double the global average.

The neonatal period is highly vulnerable for infants. In general, rural women confront many social and economic problems to keeping their children safe. Therefore, to reduce neonatal death, greater attention must be focused on this crucial period. Globally, the neonatal death has been remarkably reduced to the value of 19 deaths per 1000 live births, a 37% reduction, since 2000.

3.4.2.4 Postneonatal mortality

Postneonatal mortality is the death of children aged 29 days to 1 year. The main reasons for postneonatal deaths are malnutrition, infectious disease, troubled pregnancy, SIDS, and problems with the home environment [25].

Causes of infant mortality

About 75% of all neonatal deaths occur during the first week of life, and annually about 1 million deaths of newborn infants occur in the first 24 h. The main causes of death of children from the end of the neonatal period and through the initial 5 years of life are pneumonia, diarrhea, birth defects, and malaria. Malnutrition is particularly harmful to infants, causing severe diseases.

Timely availability of health services

Generally, the causes of infant mortality are low birth weight, congenital malformation, infectious diseases, sudden infant death syndrome, and inadequate financial condition of family.

In addition, timely vaccination with diphtheria-tetanus-acellular pertussis vaccine, haemophilus influenza type b (Hib) vaccine, hepatitis B (HepB) vaccine, inactivated polio vaccine, pneumococcal vaccine, and inactivated COVID-19 vaccine is essential for developing immunity within children. In general, the necessary vaccination should be under taken in consultation with a physician. On-time, national vaccination throughout childhood is essential because it helps provide immunity before children are exposed to potentially life-threatening diseases. Vaccines are tested to ensure that they are safe and effective for children to receive at the recommended ages.

Premature birth

Premature or preterm birth (PTB) is defined by the WHO as "babies born alive before 37 weeks of pregnancy are completed. There are sub-categories of preterm birth, based on gestational age: extremely preterm (less than 28 weeks); very preterm (28 to 32 weeks); and moderate to late preterm (32 to 37 weeks)."

Induction or caesarean birth should not be planned before 39 completed weeks unless medically indicated [26].

About 15 million babies are born too early every year, which represents more than one in 10 babies. Due to complications of preterm birth, about 1 million children die each year [27]. Many survivors of premature birth suffer from a lifetime of disability, including learning disabilities and visual and hearing problems. All over the world, premature birth has been an increasing incident resulted in causality among children under the age of 5 years.

In low income groups, due to inadequate financial conditions, about 50% of babies born at or below 32 weeks—2 months earlier than the expected time—will not survive. This is mainly due to lack of cost-effective care.

Solutions for premature birth

Varieties of reasons are responsible for premature birth. Many occur spontaneously due to family or social reasons, but some are due to early induction of labor, which may be due to either medical or nonmedical reasons. However, commonly, premature delivery occurs due to multiple pregnancies, infections, and chronic conditions such as diabetes and high blood pressure. Sometimes it is difficult to find out the real cause of premature birth.

Premature babies are more prone to chronic health problems, which may require intense healthcare in the hospital. Commonly, they suffer with infections, asthma, and feeding problems. Premature infants are also at increased risk of sudden infant death syndrome (SIDS). More than three quarters of premature babies can be saved with feasible, cost-effective care, such as essential care during childbirth and in the postnatal period for every mother and baby, provision of antenatal steroid injections (this is mainly given to strengthen the babies' lungs), kangaroo mother care (baby to be carried by mother, in close contact to body skin), frequent breastfeeding, and antibiotics to prevent newborn infections.

Consequently, it is highly necessary to provide quality healthcare to both the mother and premature baby with the help of high technology-based baby care with computerized equipment. Equipment commonly used in the NICU (neonatal intensive care unit) includes a heart or cardio respiratory monitor. This monitor displays a baby's heart and breathing rates and patterns on a screen. In addition, minor instruments equipment like blood pressure monitor, temperature meter, and pulse-oximeters should be easily accessible for premature babies and mothers.

Sudden infant death syndrome

Sudden infant death syndrome (SIDS) is an unexpected incident for newly borne babies. SIDS is the most common cause of death of infants between 2 weeks and 1 year of age. In this unclear biological process, usually during sleep, the infant stops breathing. Although the real cause is not yet clear, it is understood that this unusual biological occyrance might be due to defects in the area of an infant's brain that controls breathing and arousal from sleep. The rate of SIDS occurrence is 0.5/1000 births in the United States. The United States Center for Disease Control reports SIDS to be one of the leading causes of death in infants aged 1 month to 1 year old [28]. For this reason, the American Academy of Pediatrics recommends providing infants with safe sleeping environments, and no exposure to smoke or alcohol during pregnancy.

Congenital malformations

The World Health Organization defines congenital malformation as "structural or functional anomalies (for example, metabolic disorders) that occur during intrauterine life and can be identified prenatally, at birth, or sometimes may only be detected later in infancy, such as hearing defects."

Globally, about 295,000 newborns die within 28 days of birth every year, mainly due to congenital anomalies [29]. This biological defect leads to long-term disability, resulting in significant impacts on individuals, families, and healthcare systems. The most common congenital anomalies in newborn babies are heart defects, neural tube defects, and Down syndrome. Genetically inherited characteristics, infections, and nutritional or environmental factors are also responsible for congenital anomalies. However, some congenital anomalies can be prevented through timely vaccination, sufficient intake of folic acid, iodine fortification of staple foods or supplementation, and timely antenatal care.

The Sixty-third World Health Assembly, held at Geneva from 17 to 21 May 2010, passed a resolution to promote primary prevention and improve the health of children with congenital anomalies through data survey analysis on congenital problems of newly born babies, by giving support for research to find out or resolve problems related to congenital birth issues, and promoting international health agencies to resolve the issues at a global level.

It has been difficult to find out the specific causes of congenital anomalies but some genetic disorders and environmental factors cannot be ruled out. Congenital malformations have been noticed significantly in infant mortality. This is common to many rural areas of developing countries.

Low birth weight

Low birth weight is defined by the WHO as "weight at birth of <2500 grams (5.5 pounds)."

At a global population level, infants with low birth weight are an indicator of multifaceted public health problems resulting in long-term maternal malnutrition, ill-health, and poor healthcare in pregnancy.

In 2012, the World Health Assembly (WHA) passed a resolution on Implementation Plan on Maternal, Infant and Young Child Nutrition, which identified six global targets on priority nutrition outcomes to be achieved by 2025. After a couple of years, in 2014, the member states approved the Global Nutrition Monitoring Framework (GNMF) on Maternal, Infant and Young Child Nutrition, which included six global targets.

The targets of GNMF are: (i) to regulate the progress related to successful implementation of the six global targets to complete the goal by 2025, (ii) to find out proper implementation of the specific goal on priority basis, based on the urgency of implementation, and (iii) to keep in touch with the progress of ongoing projects to achieve targets at national level by the end of 2025 [30,31].

In general, low birth weight is due to intrauterine growth restriction, premature birth, and intake of lower quality nutrients in food due to shortcomings in family budget. Low birth weight is more common in developing countries or in least developed countries. But, due to the practice of home delivery, data on low birth weights are not properly available. Data on low birth weight (LBW) is more commonly available in rural areas of Asian countries [32–34]. Infant birth weight significantly contributes to future health problems of children and results in development of an unhealthy community. Poor maternal health and nutritional status are the main risk factors for LBW in developing countries [35,36]. Intake of a balanced diet with sufficient essential elements and energy-rich proteins may be helpful in giving birth to a healthier child.

Malnutrition

The World Health Organization defines malnutrition as "deficiencies, excesses, or imbalances in a person's intake of energy and/or nutrients" [37].

Malnutrition can be well explained under two board groups: (i) "malnutrition under nutrition," which covers stunting, wasting, underweight, and micronutrient deficiencies or insufficiencies and (ii) "overweight."

Malnutrition is a global problem. About 1.9 billion adults are under the grip of malnutrition, and 462 million are severely underweight. In addition, about 41 million children under the age of 5 years are overweight or obese, while 159 million are suffering from stunting and 50 million are wasted. In addition, about 528 million or 29% of women of reproductive age around the world are suffering from anemia due to iron deficiency. This is mainly due to rural women from developing countries being unable to afford or access a nutritious diet, with energy-rich proteins, fresh fruit and vegetables, legumes, meat, and quality milk. On the other hand, highly fatty food and high sugar and salt drinks are cheaper and more readily available, resulting in the problem of obesity in young children.

In April 2016, the United Nations General Assembly agreed to accept the proposal of "UN Decade of Action on Nutrition from 2016 to 2025." The UN Decade of Action on Nutrition is a commitment by United Nations member states to undertake 10 years of sustained and coherent implementation of policies, programs, and increased investments to eliminate malnutrition in all its forms, everywhere, leaving no one behind.

Infectious diseases

Globally, the major causes of death of children under 5 years of age are malaria, pneumonia, diarrhea, HIV, and tuberculosis. For children above 5 years, the main causes of death are noncommunicable diseases, injuries, and conflict. Despite being preventable and treatable, a significant proportion of infant death occurs due to infectious diseases. Pneumonia, diarrhea, and malaria, which can be prevented, still cause the death of many infants, and may be due to negligence on the part of parents or lack of timely treatment in the primary healthcare system. Children in the world's poorest regions are disproportionately affected, with infectious diseases particularly prevent in Sub-Saharan Africa. These trends can be preventable. UNICEF is the world's leading supplier of vaccines to developing countries as part of its commitment to child survival. UNICEF supplies vaccines to over 40% of the world's children. UNICEF, in collaboration with Gavi, Vaccine Alliance, WHO, and CEPAL are working with manufacturers and partners on the procurement of COVID-19 vaccine doses, as well as freight, logistics, and storage (Fig. 3.3).

UNICEF is able to procure and deliver for 92 low- and lower-middle-income countries. Furthermore, UNICEF also supports procurement for more than 97 upper-middle-income nations. In addition, UNICEF is also taking care of procuring and transporting immunization supplies such as syringes, safety boxes for their disposal, and cold chain equipment such as vaccine refrigerators.

Environment and infant health

Infants are extremely susceptible to different pollutants in and around their living places. Even during pregnancy, women confront various health problems due to abnormal environmental problems. Both urban and rural women are exposed to a variety of harmful pollutants caused by human activities. This contributes to diseases that account for almost 1 in 10 of all deaths of children under the age of five, and can harm the healthy development of children's brains.

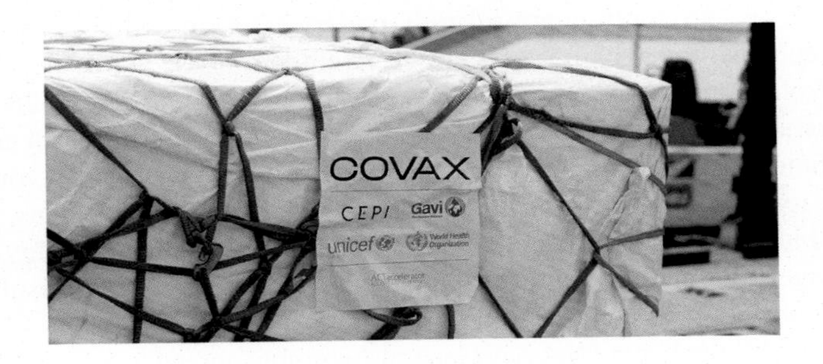

FIG. 3.3

COVAX packing, ready for dispatch by UNICEF.

Children breathe twice as quickly as adults, and take in more air relative to their body weight. Their respiratory tracts are more permeable and thus more vulnerable. Their immune systems are weaker and their brains are still developing. Particulate matters (PMs) present in the air are extremely harmful for infants. Harmful particulates can cross the blood brain barrier, which is less resistant in children, causing inflammation, damaging brain tissue, and permanently impairing cognitive development. PMs can even cross the placental barrier, injuring the developing fetus, when the mother is exposed to toxic pollution. Air pollution is consistently associated with postneonatal mortality due to respiratory effects and sudden infant death syndrome.

Carbon monoxide (CO) is a colorless and odorless gas. Presence of carbon monoxide gas in or around infants and pregnant women causes severe damage to infants due to their immature respiratory system. Children younger than 4 years and unborn babies are especially at risk of CO poising. CO can build up in a child's body and replace oxygen in his or her blood. The children's brain, organs, and tissues can be damage from lack of oxygen.

Another major pollutant is secondhand smoke, which causes detrimental mental effects in infants and fetuses. Secondhand smoke is the combination of smoke from the burning end of a cigarette and the smoke breathed out by smokers. Secondhand smoke contains hundreds of toxins and may be responsible for cancer [38–41].

Smoking during pregnancy results in more than 1000 infant deaths annually [2]. Infants exposed to second-hand smoke after birth have significantly higher risk of SIDS. Exposure to secondhand smoke causes multiple health problems in infants and children, including ear infections, respiratory system and breathing problems, and acute lower respiratory infections.

Generally, rural women working in pesticide-sprayed crop fields during pregnancy face serious health problems to both the mothers and babies during the carrying stage. Exposure to pesticides can increase the chances of having a miscarriage, a baby with birth defects, or other problems. Some pesticides may also able to pass into breast milk (Fig. 3.4).

SEven month pregnat women
working in field exposed to
pesticides

FIG. 3.4

A pregnant woman working in the rice field immediately after spray of pesticide.

Early childhood trauma

Early childhood trauma is explained as the traumatic experiences that children aged 0–6 years face. Generally, in this stage, children are not capable of expressing their anguish verbally like older children. The main causes of early childhood trauma are serious sickness, natural disasters, family violence, sudden separation from a family member, being the victim of abuse, or suffering the loss of a loved one. Early traumatic events may lead to severe consequences throughout adulthood, causing posttraumatic stress disorder, depression, or anxiety [42].

Social factors for infant mortality

The history of infant mortality is a long standing incident since many decades. It was in the early 1900s that countries around the world started to notice that there was a need for better child health care services. Social sciences defines social factors as those "under which people are grouped into a set of hierarchical social categories, the most common being the upper, middle and lower classes." The United States Children's Bureau made a survey across eight cities between 1912 and 1915, and found out that in the lower income group, infant mortality was 357% more common than for the highest income earners ($1250+). Differences between races were also observed. The incident of infant mortality rate in African-American mothers was 44% higher than average [43].

Disparities in socioeconomic factors play a key role in infant mortality. People with lower incomes, particularly those of rural areas in developing countries, are not in able to access high technology-based healthcare services during predelivery and postdelivery periods. Social factors may govern the provision of pre- and postdelivery facilities seen as being acceptable for a specific community. Developed countries, most notably the United States, have seen a divergence between those living in poverty and those who can afford advanced medical recourses, leading to an increased chance of infant mortality in the former.

Infant mortality in war

Worldwide, millions of children are victims of conflict due to countries fighting over territory, and many other reasons [44]. Over many decades, school, hospital, and healthcare workers have been brutally victimized [45,46].

Infants may become victims directly in war-affected areas, and the catastrophic impact of war discourages women in planning on having a baby. In addition, countries in the tragic grip of war and conflict suffer many other significant factors influencing infant mortality rates. Due to damage to healthcare units and transport systems, providing basic medical facilities becomes increasingly difficult. During the Yugoslav Wars in the 1990s, Bosnia experienced a 60% decrease in child immunizations. Preventable diseases can quickly become epidemic given the medical conditions during war.

Many international healthcare agencies, like the WHO and UNICEF, provide aid for basic nutrition and vaccination. Due to partial or complete paralysis of transport systems, supply chain management comes to a standstill. In such situations, the average weight of a population will drop substantially. Expecting mothers are affected even more by lack of access to food and water. During the Yugoslav Wars in Bosnia, the number of premature babies born increased and the average birth weight decreased [47].

Furthermore, forms of violence such as rape and abuse of women are frequently used as weapons of war. Women who become pregnant due to rape in war face chronic challenges in bearing a healthy child. It has been reported that women who experience sexual violence before or during pregnancy are more likely to experience infant death in their children [48–50].

Culture

Cultural inheritance social practices are also responsible for infant mortality. In many developing countries like India and Bangladesh, preference given to male over female babies causes many premature abortions and infant deaths. In developing countries such as Brazil, infant mortality rates are commonly not registered. The main cause of such practices is lack of manpower and provision for financial support.

In Ghana, there are social restrictions and prejudices against wives or newborns leaving the house. Therefore, it has been a problem to provide them health services in a timely fashion, resulting in infant mortality and even death of newly delivered mothers.

In developed countries like the United States some death in infants occurs throughout the years. The postneonatal mortality risk (28–364 days) was highest among continental Puerto Ricans compared to babies of non-Hispanic backgrounds. This is mainly due to teenage pregnancy, obesity, diabetes, and smoking [51].

In some developing countries and least developed countries, cultural beliefs may be related to the prevention of appropriate care for sick children and lead to infant mortality [50–54]. In Nigeria, even before colonial days, belief in traditional medicine and angry gods and evil spirits were major causes of illness and infant mortality [55].

3.5 Measures for prevention of infant mortality

The infant mortality rate can be reduced with the collaborative and mutual understanding of governments with various internationally recognized health agencies, nongovernment organization (NGOs), research and educational institutes, and other private bodies.

Globally, significant progress has been achieved in reducing the infant mortality rate. In 1990, about 1 in 11 children died before reaching the age five. This has been reduced to 2 in 27 children dying before reaching age five in 2019. Yearly, the global rate of infant death has been reduced to 1.9% in 2019 as compared with the figure of 3.7% in the 1990s. In spite of global progress in reduction in infant mortality rate over recent decades, an estimated 5.2 million children under age five died in 2019; more than half of those deaths occurred in Sub-Saharan Africa.

Therefore, it is high time to make progress on the followings issues to further bring down the infant mortality at a global level.

3.5.1 Amendments to policy

Governments can reduce infant mortality by addressing the combined need for education (such as university education), nutrition, and access to basic maternal and infant health services. Maximum emphasis should be given to improving access to rural women's health care for those are exposed to potential risk during the pre- and postpregnancy period. Most of the infant death in rural communities is due to low birth weights and contracting pneumonia, which can be improved by providing air quality hygiene care to prevent infant mortality. In this connection, the government should extend financial support for developing home-based technology to chlorinate, filter, and solar disinfect organic water pollution, which will be helpful in reducing diarrhea in children significantly [56–58].

Improvements in providing quality food and sanitation aids can reduce infant mortality, as seen in the United States' vulnerable rural community populations, including African Americans [59].

As reported by UNICEF, hand washing with soap before eating and after using the toilet can save more lives of children by cutting deaths from diarrhea and acute respiratory infection [60]. By controlling the low birth weight deliveries throughout the population, the United States has achieved a reduction in infant mortality rates on a regional population level [51].

At government level, while planning budgets, adequate financial provision should be kept in healthcare planning so that necessary healthcare providers are available at rural community level. With the addition of one physician per 10,000 people, there is potential for 7.08 fewer infant death per 10,000 [61].

3.5.2 Prenatal care and infant mortality

The healthcare provider, especially in rural areas, should raise awareness among pregnant women of the importance of receiving a regular healthcare check at a nearby primary health center, to update information on the baby's chances of being delivered in safe conditions and surviving. In addition, the biological importance of a balanced diet with folic acid, B_9 (or supplementary of folic acid), should be carefully explained to local rural women. Folic acid can reduce infant mortality [62]. In many countries there is provision of mandatory folic acid supplementation in the diet of pregnant women to avoid spina bifida, a birth defect in newborns [63,64].

Pregnant women with an alcohol or smoking habit and tobacco use suffer from birth defects and low birth weight birth, respectively, which ultimately leads to infant death [65,66].

3.5.3 Nutrition

Improving maternal, infant, and young child quality nutrition expands opportunities for every child to reach his or her full potential. Proper nutrient deficiency during pregnancy may lead to stunting and wasting after birth. Stunting is mainly due to chronic or recurrent under nutrition in utero and early childhood. Wasting is a life-threading condition mainly due to poor nutrition and less energetic proteins. Due to wasting, infants and children under 5 years of age face malnutrition and week immunity. About half of deaths in children under 5 years are attributable to under nutrition. The American Academy of Pediatrics recommends exclusive breastfeeding of infants for the first 6 months of life, following by a combination of breastfeeding and other sources of food through the next 6 months of life, up to 1 year of age [67].

3.5.4 Vaccinations

Vaccination is a vital practice to develop immunity, and a preventive measure to save infants and children from various types of bacterial or virus infection. In order to prevent infant mortality, the Centers for Disease Control and Prevention (CDC) recommends the following vaccinations to infants from 1 month to 1 year of age [68]:

> *Hepatitis B (HeB), rotanirus (RV), haemophilus influenza type B (HIB), pneumococcal conjugate (PCV13), (IPV < 18 yrs), influenza, varicella, measles, mumps, rubella (MMR), and diphtheria, tetanus, acellular pertussis (DT TapP < 7yrs) inactivated poliovirus.*

Each of these vaccinations are given at particular age ranges depending on the vaccination and are required to be done in a series of one to three doses over time, depending on the vaccination.

3.5.5 Socioeconomic factors

A range of social factors are also responsible for bringing down the infant mortality rate. Educating mothers, especially rural women, is essential. This can be carried out by the local healthcare work force to provide them basic information on family planning, improvement on children's health, and lowering infant mortality rate. For example, in Sub-Saharan Africa an increase in women's education brought a reduction in infant mortality of about 35% [69]. Educational attainment and public health campaigns provide the knowledge and means to practice better habits and lead to better outcomes against infant mortality rates.

The GDP acts as a best indicator for the general state of the economy. Mostly, in recession, two quarters of negative economic growth occur. The impacts are realized at individual level on the basis of poverty and unequal effects on different socioeconomic groups. GDP per capita in recession periods has an impact on the general health of the economy that is available universally, and especially for developing countries under the constraints of poverty. The link between individual income and health outcomes is well studied. The poorer countries generally have inferior health status. Recession in the economy may have impact on health services in a variety of ways (Fig. 3.5).

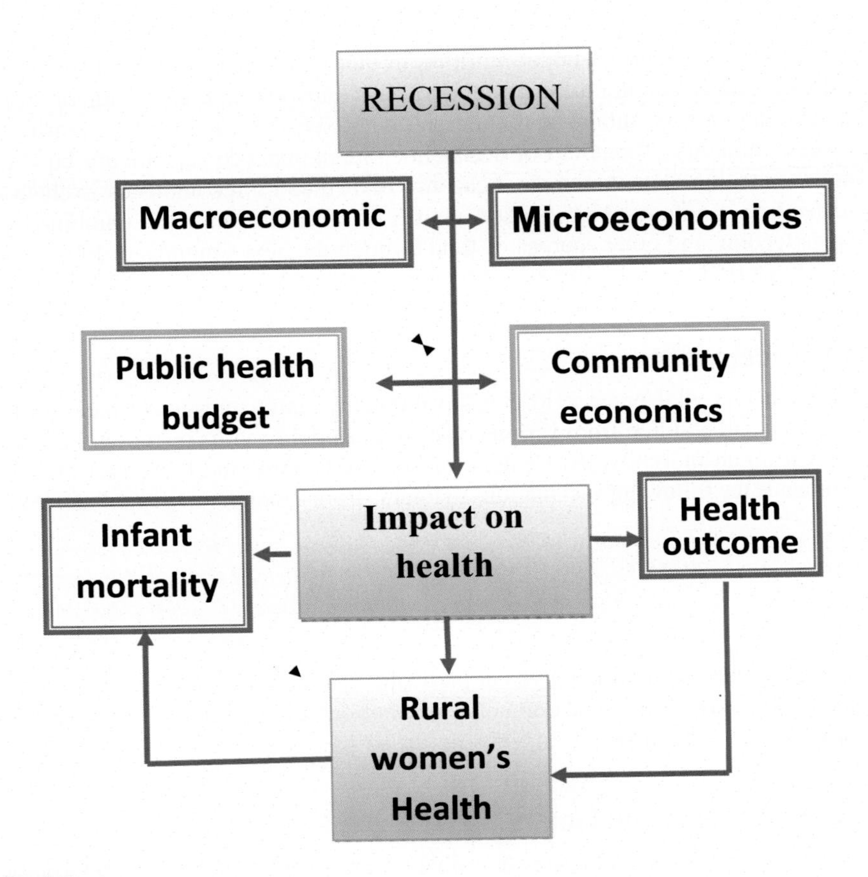

FIG. 3.5

Various routes through which recession may impact on health outcomes. The numbers in this figure link to textual references.

3.6 Difference in IMR expression

Expression of infant mortality varies widely between countries based on how they define a live birth and how many premature infant are born in the country. The World Health Organization (WHO) defines a live birth as: "any infant born demonstrating independent signs of life, including breathing, heartbeat, umbilical cord pulsation or definite movement of voluntary muscles" [70].

Austria uses this definition without any further amendment [71], but in Germany it is expressed with slight modification: muscle movement is not considered to be a sign of life [72]. Many other countries, including certain European states (e.g., France), only count as live birth cases where an infant breaths at birth, which makes their rates of IMR numbers somewhat lower and increases their rates of perinatal mortality [73]. The parameter requirement for live birth in the Czech Republic and Bulgaria is still higher [74].

Even with the increase in and application of new information technology, there are some developing countries and least developed countries that have the unusual practice of not registering infant mortality cases.

UNICEF, in coordination with the WHO, the World Bank (WB), and the United Nations Statistic Division (UNSD), compile infant mortality data from all over the world. The method of estimation is obtained either from standard reports, direct estimation from existing data bank, or on the basis of official annual data survey by UNICEF.

3.7 Puberty

Puberty is a biological process of life that generally occurs in girls in the period of 10–11 years and is completed between ages 15 and 17 [75–77].

The onset of puberty and adulthood (teenage) is prone to misshapen, if proper attention is not given by parents in the family and teachers in the school. Mostly, in rural areas of developing countries, teenage girls face a variety of problems due to poverty, hunger, lack of education, and inheritance socially exclusive problems. The following are a few such problems that are commonly faced by rural girls.

3.7.1 Girl's education in rural areas

The current UNESCO report says about 132 million girls are not seen school in their life. This figure is about 32 million of primary school age, 30 million of lower-secondary school age, and 67 million of upper-secondary age [5].

In each developing country, strategically, priority should be given to girl's education. Better education of women and girls will be helpful in maintaining a healthy family life and be helpful in developing their respective younger generation with sound health and better quality of life. In addition, education will provide an opportunity to rural girls to understand the outer world, labor market, community socioeconomic problems, primary skills for self-dependence in life, and contribute to knowledge on agriculture and development to their respective elderly parents for better agriculture practices. However, unfortunately, governments pay less attention and budgetary provision for rural education development. Consequently, multiple hurdles are faced by rural girls to get a proper education.

3.7.2 Poverty

Lack of basic infrastructure for primary schools, low numbers of education providers, deficiency of learning materials, poor sanitation facilities, and no transport services make learning difficult for girls in rural areas of developing counties. Currently, access to learning is the biggest challenge to provide basic education to children, especially for rural girls, for making them an ideal citizen. UNICEF reports that about 11% of primary school-aged children and 20% of lower-secondary-aged children have not seen school. Besides, an estimated 617 million children and adolescents did not receive the opportunity for minimum proficiency level in reading even though two thirds of them are in school. For girls residing in rural areas of developing countries, only 49% have a limited education opportunity. Rural girls are often excluded from primary education due to gender priority.

3.7.3 UNICEF education

Due to the lack of skills for lifelong learning, children, especially girls, have deficiencies in self-confidence for earning. Furthermore, due to lack of primary knowledge in healthcare, they are more likely to suffer adverse health outcomes. The sudden outbreak of COVID-19 has challenged the global community for continuation of education. To meet the deficit in providing education, digital learning has become an essential service, globally. However, to extend world-class digital learning to rural areas has been a great challenge to governments of developing countries. Rural children who receive primary education under trees or in poorly hatched classrooms cannot even dream of digital learning (Fig. 3.6).

This great disparity in providing learning facilities to rural children, as compared to urban children, not only effects rural children at an individual level, but also restricts rural community development, which feeds manpower, food, and unskilled labor for urban development and management. So, UNICEF has a target to develop primary education at a rural level by promoting the following.

3.7.3.1 Access to education

Gender equality is unavoidable and essential to provide quality education from early childhood to adolescence, especially for rural girls under extreme stress of poverty. According to a joint report provided by UNICEF and the International Telecommunication Union (ITU), about 1.3 billion children aged 3–17 years old lack an internet connection in their home [78].

FIG. 3.6

Rural children learning under trees of a village, India.

Absence of internet services in rural areas not only restricts the individual student and younger generation, but also prevents them from accessing the outer world to understand science, technology, and literature development and their role in the progress of community and nation. It isolates them from the world. Outbreaks of COVID-19 have meant that nearly a quarter of a billion students at a global level have come to rely on virtual learning. For rural students, due to lack of internet services, education has become out of reach. A big disparity exists in internet services between urban schools and village schools (Fig. 3.7).

About 58% of school-age children from upper class families have internet connections at home, compared with only 16% from lower-middle class families. The same disparity is noticed across the country level in lower income groups. Less than one in 20 school-age children from lower income countries have an internet connection at home. A large gap exists between developed and developing countries in use of modern internet facilities. About 1.3 billion school-age children, mostly from lower-income and rural areas, are at risk of missing out on education because they

FIG. 3.7

Internet disparity in Internet service at different level of school located in (A) urban and (B) rural areas.

lack access to the internet at home. Geographical disparities are also an important issue related to providing internet services. Worldwide, about 60% of students from urban areas are unable to access the internet at home, compared to about three-quarters of school-age children in rural households, as noticed in Sub-Saharan Africa and South Asia, where 9 in 10 children are deprived of a network connection.

3.7.3.2 Quality learning and skill development

Quality education is a vital tool for enhancing quality of life, developing awareness and capability, and increasing self confidence in learners, especially those residing in rural areas. Quality education helps rural communities to lead healthier and more sustainable lives. In most rural communities of developing countries the family is male dominated. So, women feel helpless in every area of life, especially in giving education to girls. So, it is the foremost duty of the government to provide special budgetary provision for the development of education facilities in rural areas. An education system in rural communities opens the scope of learning capacity and helps girls to make informed decisions about their agriculture farms, guiding their respective parents to improve productivity, not to expose them in rice or wheat fields immediately after pesticide spraying, to think for themselves in obtaining skill education, which will be applicable in their family life. A rural educated girl or woman can understand policies, procedure, rights, duties, government schemes, legislation, available benefits, and protection laws.

Provision of quality education and skills to rural girls will be helpful to manage the family, agriculture, and other community activities when the male family members migrate to urban areas for better job opportunities.

Currently, about 69% of India's population lives in rural areas. Out of 135 crore population of India, women constitute 48% of the total rural area. India is the second most populous country next to China. Quality education to rural girls and women can enhance the rural quality of life and control further population increase.

3.7.3.3 Emergencies and fragile contexts

Education can play a significant role in developing accelerated progress for people living in fragile states. Rural education, especially to girls and women, will help them to reduce the security gap, capacity gap, and legitimacy gap. Generally, rural girls and women are not provided with adequate protection. Rural girls and women also face a capacity gap due to lack of service opportunity. Due to lack of education, rural girls and women do not have knowledge of legal protection and security.

3.7.4 Violence

It is the right of children to receive education at school level free from fear. With safe and secure education, especially in rural areas of developing countries, children can develop good friendships among themselves for better learning and understanding. But, due to lack of school security and teacher's care, female students face violence, mainly due to harassment, verbal abuse, sexual abuse, and exploitation. This is still happening in schools located in rural areas.

School children victimized by violence may experience physical injury, sexually transmitted infections, depression, anxiety, posttraumatic stress disorder (PTSD) and even suicidal thoughts.

UNICEF, in association with governments and nongovernment organizations (NGOs), is trying to provide protection measures to school students with special attention for rural schoolchildren.

3.7.5 Child marriage

Child marriage is an evil practice, mostly occurring between girls under the age of 18 with an older man, and are rooted in gender inequality. However, child marriage has reduced globally in recent years. A remarkable decline has been seen in South Asia in the last decade. Girls' marriage before the age of 18 has been reduced to 30%, in large part due to progress in India. Increasing rates of girls' education, proactive government investigations into adolescent girls, and strong public messaging on the illegality of child marriage and the harm it causes are the main reasons for the reduction in child marriage in rural India. Forced child marriage causes a girl lifelong suffering due to not getting opportunities for learning. The worst consequence of child marriage is earlier pregnancy, leading to complications in child delivery, which often risks the life of both mother and infant at the early stage of birth.

The sustainable development goals (SDGs) require the reduction of the rate of child marriage. As reported by UNICEF, about 12 million child marriages take place each year in rural areas of developing countries as compared to an expected value of 25 million around the world. However, to stop the practice by the end of 2030, the implementation of a plan to eradicate child marriage is in progress. Globally, about 650 million women alive today were married as children. Meanwhile, South Asia has shown significant progress in reduction of child marriage. Currently, the burden of child marriage has been shifted to Sub-Saharan Africa.

The Sustainable Development Goals (SDGs) have a target to completely eradicate child marriage by 2030. The SDGs have 17 goals targeted for global development priorities between now and 2030. Target 5.3 aims to eliminate all harmful practices, including child marriage and female genital mutilation, by the end of 2030, and 193 countries have agreed to end child marriage by 2030 under the SDGs.

The root causes of child marriage are poverty, gender inequality, insecurity, and the lack of economic and social opportunities for girls. So the main reason for applying the SDGs around the world, especially focusing in developing countries, is to bring an end to child marriage.

3.7.5.1 Impact of war

The outbreak of war overwhelms the life of a country. Generally, children are dependent on parental care and attention. Their lives can be mercilessly disrupted by violence and the loss of parents on the war front. Children may lose opportunities for education during war, and can be forced to move to refugee or temporary shelter sites where taking care of life is uncertain. They have to wait for an uncertain time under miserable circumstances for normal life to resume, if it ever does. Children who survive may be disabled during war with loss of sight, limbs, and other physical inabilities, causing loss of opportunity for schooling and social life. Girls victimized by rape may be marginalized by society and lose the opportunity for marriage. After the end of war, return to a normal life is a big challenge within their society.

Uncounted numbers of innocent children lose their lives in wars every year [79]. They die as civilians caught in the violence of war or are directly. Injuries to children are differentiated on the basis of attack by different types of weapons. Explosion of land mines during war is more injurious to children than the adults [80,81]. Due to partial or serious injuries, millions of children are disabled, and rehabilitation of them after war is a big issue and challenge for governments.

3.8 Reproductive health of rural women

Reproductive health of women refers to sexual health (hygiene) and addresses the reproductive process, functions, and system at all stage of life.

Rural women confront numerous constraints in accessing affordable, adequate health services (e.g., clinics, hospitals, reproductive health/family planning, and counseling) during the reproductive stage of their life cycle. Inadequate transport facilities and restrictions on their mobility create unfavorable conditions for them to access health services in a timely fashion.

3.8.1 Right to reproductive health

The right to sexual and reproductive health forms part of the "right to the highest attainable standard of physical and mental health." Sexual and reproductive health and rights (SRHR) are complementary to universal health coverage (UHC) (Fig. 3.8).

Under the Sustainable Development Goals (SDGs), as part of the process and progress of UHC, countries should also consider how the social significance of SRHR will be helpful in a compatible rural health community. Due to its cost-effective nature, SRHR can be affordable by most developing countries. Several countries have made significant progress in the implementation of SRHR interventions [82].

In this connection, the International Conference on Population and Development (ICPD) was held in 2019, and 179 governments agreed that human rights, including reproductive rights, were fundamental to development and population concerns. So, all countries should adopt reproductive healthcare practices like voluntary family planning/contraception and safe pregnancy and childbirth services. The most remarkable feature of the ICPD Program of Action was that it linked reproductive rights to human rights. So, family planning was no longer merely aimed at reducing fertility and decreasing population growth but was seen as a means to empower women and to promote rights and choices in relation to reproduction. The benefit was not limited to women's empowerment but helpful in preventing unsafe abortion,

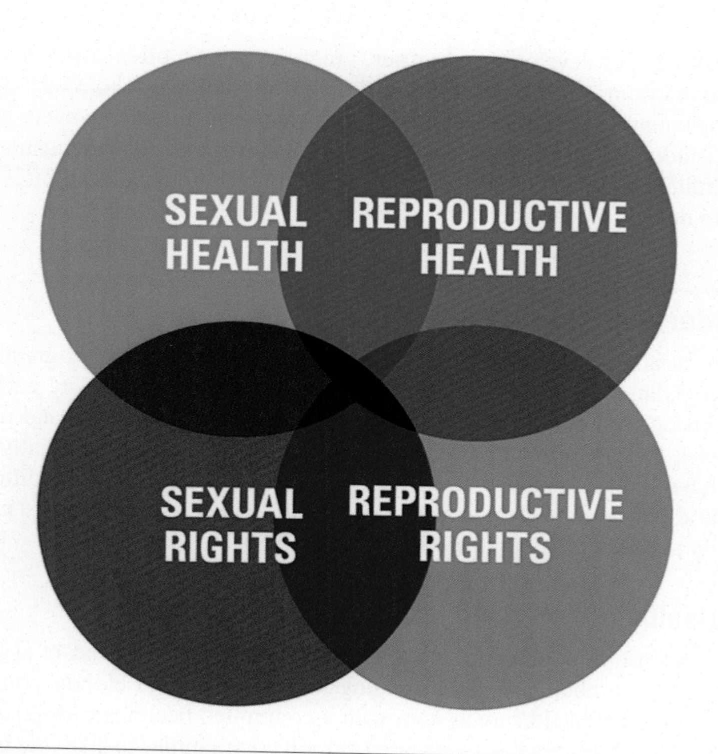

FIG. 3.8

Toward a compressive SRHR concept.

improving women's capacity to avoid or receive treatment for STIs, including HIV infection, and the care that she receives during pregnancy and childbirth.

3.8.2 Reproductive health in developing countries

People from developing countries are at great risk from sexually transmitted diseases (STDs). About 585,000 of the women worldwide who die due to pregnancy-related complications annually are from Africa, Asia, and Latin America. Many of these deaths could have been prevented but are not, due to inadequate financial budgets, lack of health service providers, and unstable government, and insufficient resources devoted to reproductive health care and family planning.

STDs are chronic diseases and are widespread throughout developing countries. The most severe consequences include a high likelihood of HIV transmission, infertility, death from ectopic pregnancy, or permanent disabilities for newborns. In this connection, high risk groups such as prostitutes and clients are dangerous sources for HIV and other STDs. Therefore, it is essential that pregnant women should be routinely screened for syphilis, and eye medicine should be administered to newborns to prevent blindness that may result from an STD transmitted from mother to child.

In rural areas of developing countries, pregnant women often suffer from hemorrhaging and pronged labor, and this is difficult to predict, as about 60% births in developing countries take place outside of health facilities. In order to overcome such problems, older female family members and midwives or birth attendants should be well trained to understand life-threatening complications so that cases can be referred to nearby primary healthcare centers for further care.

3.9 Elderly years and rural women

Currently, rural and remote areas of developing countries are facing greater problems of population aging than urban areas. Lower population density and scatted distribution of rural communities (villages) has presented a difficult and more expensive task for governments. Consequently, rural populations, especially elderly women, have less access to services under their poor socioeconomic conditions. The elderly rural population face greater risk of social isolation due to personal problems in mobility and lack of support and healthcare [83,84].

3.9.1 Healthcare services for the elderly population

With the increase in population aging, the present model of healthcare system and services will be a challenge for extending health services to elderly populations. A well-coordinated healthcare system with a committed healthcare work force and specialized physicians need to be deputed in rural communities to take care of the aging population. Additional residential facilities should be provided to the healthcare worker for their security and children's education. Informal care provided by family and friends will also need to be supported to allow people with functional limitations to stay at home as long as possible. The community healthcare head should be empowered to select a potential workforce and train them for the aging population. For a better service, experienced doctors from community healthcare should be deputed in rural areas for timely vaccination programs and health measures to protect the aging population from communicable diseases. The community healthcare doctor should be entrusted with financial power for home care services to elderly women on a need basis.

3.9.1.1 Health workforce

Due to variable climatic factors and increases in environmental contamination, health problems in the aging population have been increasing. Older adults, especially women, have different healthcare need than younger groups, and this will affect the demands placed on the healthcare system in the future. An aging population is more prone to suffer from chronic illnesses like cancer, heart disease, and diabetes than younger people. In order to provide continuing services for an aging population the health workforce is aging in many countries. For example, in Denmark, France, Iceland, Norway, and Sweden, the average age of nurses employed today is 41–45 years.

3.9.1.2 Vaccination and infectious disease prevention

Vaccine coverage rates of elderly old people in developing countries are relatively poor as compared to their counterparts in urban areas. Even in some parts of Europe, vaccination of the aging population has reduced due to an inadequate health work force. The COVID-19 pandemic drastically grabbed the global population. In many developing countries, especially older people of rural areas are facing the most threats and challenges at this time. Although all age groups are at risk of contracting COVID-19, elderly people are at greater risk due to their lower immune capacity at old age.

In rural areas of developing countries, a number of adverse factors like poor hygienic conditions, nonavailability of quality drinking water, and just manageable livelihood conditions are responsible for serious health risks to elderly old people. Generally, the elderly people in rural areas are prone to pneumonia, influenza, tuberculosis, bacteria, and nosocomial (hospital-acquired) infection [85–87].

Timely vaccination of elderly people reduces severe illness and complications.

3.9.1.3 Long-term care

The World Health Organization defines long-term care as: "the system of activities undertaken by informal caregivers (family, friends and/or neighbours) and/or professionals (health and social services) to ensure that a person who is not fully capable of self-care can maintain the highest possible quality of life, according to his or her individual preferences, with the greatest possible degree of independence, autonomy, participation, personal fulfilment, and human dignity" [88].

This type of care can easily and conveniently be provided by formal and informal support systems. The latter covers a broad range of community services, which include public health, primary care, and rehabilitation services.

The nature of long-term care varies widely on the basis of a country's financial status and the type of sustainable government. For example the share of people aged 65 and older receiving services in institutions ranges from less than 1% in Poland and the Russian Federation to 9% in Iceland. The share of those receiving publicly funded services at home ranges from very small in many countries in Eastern Europe to 25% in Denmark. In almost all countries, most (often close to 80%) of those receiving long-term care are aged 80 and above.

3.9.1.4 Elderly care at home

Eldercare means fulfillment of the daily requirement of elderly people at home. This service includes assisted living, adult day care, long-term care, and residential care. Due to the variable nature of elderly people, including some of an abnormal nature, it is necessary to have patience and understanding to serve elderly people. Elderly care emphasizes the social and personal requirements of senior citizens who wish to age with dignity while needing assistance with daily activities and healthcare. This concept is applicable in families in urban areas of both developed and developing countries. However, in rural areas, to stay at home for elderly old people is most often inappropriate, due to poverty and the financial condition of the family. Generally, women live longer than men. In such situations, after the death of a husband spending time alone for a female can be very hard.

India is the second most populous country, next to China. About 18 million elderly in India are homeless. Senior citizens living in poverty face abandonment by their own families as they cannot earn an income. Many times they are left with no choice but to beg to survive. Many elderly are left alone after their children move to the cities in search of a better livelihood. This situation can prevail both in developed and developing countries. So, in order overcome such problems, many nongovernment organizations and private bodies, partly supported by governments, are putting effort into develop sheltered homes for homeless people to live out their remaining life under favorable circumstance.

3.9.2 Distribution of the elderly population

In developed countries like the United States, the distribution of the older population—those over 65 years—is about 46.2 million, out of which 10.6 million are living in areas designated as rural [89].

The elderly rural population needs specialized medical and rehabilitation services and conventional transport services for moving to nearby primary health centers for regular checkups. It has been predicted that in the next few decades, the older adult population will have an unprecedented impact on the US healthcare system, especially with reference to supply and demand for healthcare workers. Due to the regular aging problem, the large number of retirees may affect the management of health services. In addition, older adults consume a disproportionately large share of American healthcare services, so demand for health services will grow. The aging population may affect the efficacy of the healthcare work force. There were 101.1 million older people (>65 years) living in Europe at the start of 2018, and this value is about one fifth (19.7%) of the total population. It has been predicted that the number of elderly will increase to 149.2 million inhabitants by the end of 2050. This value is equivalent to 28.5% of the total population in 2050.

Just like other countries, population aging in Africa is in a steady growth pattern. But, policy actions on ageing in Africa are further complicated by rapidly changing environment in which older persons live. The great majority of the older persons live in rural areas where social infrastructure is scanty. So, it is great part of the failure of government to take care of elderly people. A decade back, the percentage of people aged 65 in Africa had grown to 3.6% in 2010 from 3.3% in 2000. It has been predicted that the elderly population will reach 4.5% of the continent's population by 2030, and it will be almost 10% by 2050. The significant increase in the elderly population has been noticed in middle-income African countries, such as Mauritius, South Africa, Egypt, Morocco, and Tunisia. Other countries such as Libya, Botswana, Zimbabwe, and Djibouti have also witnessed fast growth among the elderly population. However, unlike many developed countries, African nations are in general not best equipped to deal with the rise in the numbers of older people. One of the major problems is lack of pension schemes for government workers. The budgetary provision for healthcare by most governments is extremely poor. Many elderly people in Africa are burdened with children due to HIV/Aids related death of parents. UNICEF statistic show that more than 50% of orphans in Africa live with their grandparents, many on limited and uncertain incomes.

The overall health problem with elderly people is critical. Most of the African aging population suffers from long-term chronic conditions such as heart disease, cancer, respiratory disorders, and dementia. For elderly women, age and gender discrimination is a major concern that often leads to disempowerment and can result in poor health outcomes, victimization, and even death. Estimated at 43 million in 2010, the population of elderly people in Sub-Saharan Africa is projected to reach 67 million by 2025 and 163 million by 2050. In most of the African nations, healthcare related to HIV/AIDS prevention measures are restricted to younger people. Older adults are often given little information on healthy aging. Chronic poverty is a major risk factor for the wellbeing of older people.

3.9.3 Causes of population aging

One of the basic causes of population aging is the long-term fall in fertility rates and increase in life expectancy.

In higher-income groups, life expectancy has increased to 60 years in the last few decades. Reduction in the use of tobacco and decreases in the rate of cardiovascular disease mortality has contributed to a fall in male mortality since 1980.

Population aging is also a consequence of the demographic transition from higher to lower levels of fertility and mortality. In this connection, a longitudinal nationally representative household survey was conducted in China, Ghana, India, Mexico, Russia, and South Africa, with a sample size of over 40,000 respondents by Global Aging and Adult Health (SAGE) of the World Health Organization to understand the trend in fertility and mortality. The respondents for this survey were 50 years and older with a smaller, comparative cohort of adults aged 18–49 years.

The survey was further supported by eight health and demographic surveillance sites (HDSS) in Bangladesh, Ghana, India, Indonesia, Kenya, South Africa, Tanzania, and Vietnam with an additional combined sample size of over 45,000 respondents. The Collaborative Research on Aging in Europe (COURAGE) project has also used SAGE methods and instruments to collect data in Finland, Poland, and Spain [90].

The overall aim of SAGE is to improve the empirical understanding of the health and wellbeing of older adults. On the basis of firsthand information available in the survey report, it is understood that significant declines in health over the life span with female and poorer respondents was due to worse health at all ages.

The initial survey was completed in 2004, followed by a first survey in 2010 and second SAGE survey in 2015. In these surveys, the biomarker component of SAGE includes performance tests and the collection and storage of dried blood spots (DBS) samples from 40,000 respondents. The DBS were assayed for hemoglobin, glycosylated hemoglobin (HbA1c), high sensitivity C-reactive protein (hsCRP), Epstein Barr virus (FBV), and HIV. For future SAGE surveys, DNA analysis is included.

Other factors responsible for population aging include advances in public health and medical technology, development of awareness on benefits linked to a healthy lifestyle, and increased healthcare facilities in SDGs programs.

References

[1] Constitution of World Health Organisation. https://www.who.int>-constitution_en; 1946.

[2] UN General Assembly. Transforming our world: the 2030 agenda for sustainable development; 2015 [21 October. UN Doc. A/RES/70/1].

[3] Ronsmans C, Graham WJ. Maternal survival 1—maternal mortality: who, when, where, and why. Lancet 2006;368:1189–200. https://doi.org/10.1016/S0140-6736(06)69380-X.

[4] de Bernis L. Maternal mortality in developing countries: what strategies to adopt? Medecine Tropicale 2003;63:391–9.

[5] Campbell OMR, Graham WJ. Maternal survival 2—strategies for reducing maternal mortality: getting on with what works. Lancet 2006;368:1284–99. https://doi.org/10.1016/S0140-6736(06)69381-1.

[6] Paxton A, Maine D, Freedman L, Fry D, Lobis S. The evidence for emergency obstetric care. Int J Gynaecol Obstet 2005;88:181–93. https://doi.org/10.1016/j.ijgo.2004.11.026.

[7] Anon. International technical guidance on sexuality education: an evidence-informed approach (PDF). vol 22. Paris: UNESCO; 2018, ISBN:978-92-3-100259-5.

[8] Pallitto CC, García-Moreno C, Jansen HA, Heise L, Ellsberg M, Watts C. WHO multi-country study on women's health and domestic violence. Int J Gynaecol Obstet 2013 Jan;120(1):3–9.

[9] Garcia-Moreno C, Watts C. Violence against women: an urgent public health priority. Bull World Health Organ 2011;89(1):2.

[10] Coker AL, Davis KE, Arias I, et al. Physical and mental health effects of intimate partner violence for men and women. Am J Prev Med 2002;23:260–8.

[11] Macaluso M, Wright-Schnapp TJ, Chandra A, Johnson R, Satterwhite CL, Pulver A, et al. A public health focus on infertility prevention, detection, and management. Fertil Steril 2010;93(1):16.e1–16.e10.

[12] Trent M. Status of adolescent pelvic inflammatory disease management in the United States. Curr Opin Obstet Gynecol 2013;25(5):350–6. https://doi.org/10.1097/GCO.0b013e328364ea79.

[13] Janisch CP, Schubert A. WHO special program of research, development and research training in human reproduction (HRP). Geburtshilfe Frauenheilkd 1991;51(1):9–14.

[14] World Health Organisation. Ageing and health; 2018. https://www.who.int>Newsroom>Fact sheets>Detail.

[15] Farr SL, Cooper GS, Cai J, Savitz DA, Sandler DP. Pesticide use and menstrual cycle characteristics among premenopausal women in the agricultural health study. Am J Epidemiol 2004;160:1194–204.

[16] Michon S. Disturbances of menstruation in women working in an atmosphere polluted with aromatic hydrocarbons. Polski Lekarski 1965;20:1648–9.

[17] Larsen SB, Giwercman A, Spano M, Bonde JP. A longitudinal study of semen quality in pesticide spraying Danish farmers. The ASCLEPIOS Study Group. Reprod Toxicol 1998;12:581–9.

[18] Euro-Peristat. European perinatal health report: Core indicators of the health and care of pregnant women and babies in Europe in 2015. November 2018; 2018.

[19] WHO. Maternal and perinatal health; 2020. www.who.int.

[20] Richardus JH, Graafmans WC, Verloove-Vanhorick SP, Mackenbach JP. The perinatal mortality rate as an indicator of quality of care in international comparisons. Med Care January 1998;36(1):54–66.

[21] GBD 2013 Mortality and Causes of Death, Collaborators. Global, regional, and national age-sex specific all-cause and cause-specific mortality for 240 causes of death, 1990–2013: a systematic analysis for the global burden of disease study 2013. Lancet 2014;385(9963):117–71.

[22] World Health Organization. Infant mortality; 2018. https://www.who.int>Topics>indicator.Groups.

[23] Norton M. New evidence on birth spacing: promising findings for improving newborn, infant, child, and maternal health. Int J Gynaecol Obstet April 2005;89(Suppl 1):S1–6.

[24] World Health Organization. Newborn: improving survival and well-being; 2020. https://www.who.int>Newsroom>Factsheet>Detail.

[25] World Health Organization (WHO). Child health: health topics. Geneva: WHO; 2016.

[26] World Health Organization. Preterm birth; 2018. http://www.who.int>Newsroom>Fact sheet>Detail.

[27] Liu L, Oza S, Hogan D, Chu Y, Perin J, Zhu J, et al. Global, regional, and national causes of under-5 mortality in 2000-15: an updated systematic analysis with implications for the sustainable development goals. Lancet 2016;388(10063):3027–35.

[28] Mathews TJ, MacDorman MF. Infant mortality statistics from the 2009 period linked birth/infant death data set (PDF). Natl Vital Stat Rep January 2013;61(8):1–27.

[29] World Health Organization. Congenital anomalies; 2020. https://www.who.int>Fact sheets>Detail.

[30] UNICEF-WHO. Joint database on low birth weight., 2019, http://data.unicef.org/nutrition/low-birthweight;, https://www.who.int/nutgrowthdb/lbw-estimates.

[31] WHO, Global Health Observatory (GHO). Data repository. Low birth weight, prevalence (%) (Child malnutrition)., 2019, http://apps.who.int/gho/data/view.main.LBWCOUNTRYv.

[32] Anon. Report of the secretary-general. Sustainable development goals report. United Nations Sustainable Development; Sept 7 2019. https://www.un.org/sustainabledevelopment/progress-report/. [Accessed 15 January 2020].

[33] Institute for Public Health; National Institutes of Health; Ministry of Health Malaysia. National health and morbidity survey: maternal and child health, vol II: findings. 2016; 2016.

[34] Bachok N, Omar S. The effect of second-hand smoke exposure during pregnancy on the newborn weight in Malaysia. Malays J Med Sci 2014;21:44–53.

[35] Usha R. Nutrition and low birth weight: from research to practice. Am J Clin Nutr 2004;79(1):17–21. https://doi.org/10.1093/ajcn/79.1.17.

[36] Katz J, Lee AC, Kozuki N, Lawn JE, Cousens S, Blencowe H, et al. Mortality risk in preterm and small-for-gestational-age infants in low-income and middle-income countries: a pooled country analysis. Lancet 2013;382(9890):417–25.

[37] World Health Organisation. Malnutrition; 2020. https://who.int>news-room>q-a-detail>malnutriti.

[38] U.S. Department of Health and Human Services. The health consequences of smoking—50 years of progress: a report of the surgeon general. Atlanta: U.S. Department of Health and Human Services, Centers for Disease Control and Prevention, National Center for Chronic Disease Prevention and Health Promotion, Office on Smoking and Health; 2014.

[39] U.S. Department of Health and Human Services. A report of the surgeon general: how tobacco smoke causes disease: what it means to you. Atlanta: U.S. Department of Health and Human Services, Centers for Disease Control and Prevention, National Center for Chronic Disease Prevention and Health Promotion, Office on Smoking and Health; 2010.
[40] U.S. Department of Health and Human Services. The health consequences of involuntary exposure to tobacco smoke: a report of the surgeon general. Atlanta: U.S. Department of Health and Human Services, Centers for Disease Control and Prevention, National Center for Chronic Disease Prevention and Health Promotion, Office on Smoking and Health; 2006.
[41] Huang J, King BA, Babb SD, Xu X, Hallett C, Hopkins M. Sociodemographic disparities in local smoke-free law coverage in 10 states. Am J Public Health 2015;105(9):1806–13.
[42] Copeland WE, Keeler G, Angold A, Costello EJ. Traumatic events and posttraumatic stress in childhood. Arch Gen Psychiatry 2007;64(5):577–84.
[43] Haines MR. Inequality and infant and childhood mortality in the United States in the twentieth century (PDF). Explorations in Economic History 2011;48(3):418–28.
[44] UNICEF. More than 1 in 10 children living in countries and areas affected by armed conflict. New York: United States Fund for UNICEF; 2015.
[45] Education Under Attack. Global coalition to protect education from attack; 2014.
[46] Anon. Report on attacks on health Care in Emergencies. Geneva: World Health Organization; 2016.
[47] Krug E. World report on violence and health. Geneva: Geneva WHO; 2002.
[48] Asling-Monemi K, Peña R, Ellsberg MC, Persson LA. Violence against women increases the risk of infant and child mortality: a case-referent study in Nicaragua. Bull World Health Organ 2003;81(1):10–6.
[49] Emenike E, Lawoko S, Dalal K. Intimate partner violence and reproductive health of women in Kenya. Int Nurs Rev March 2008;55(1):97–102.
[50] Jejeebhoy SJ. Associations between wife-beating and fetal and infant death: impressions from a survey in rural India. Stud Fam Plann September 1998;29(3):300–8.
[51] MacDorman MF, Mathews TJ. The challenge of infant mortality: have we reached a plateau? Public Health Rep 2009;124(5):670–81.
[52] Ariyo O, Ozodiegwu ID, Doctor HV. The influence of the social and cultural environment on maternal mortality in Nigeria: evidence from the 2013 demographic and health survey. PLoS One 2017;12(12), e0190285.
[53] Fayehun O, Obafemi O. Ethnic differentials in childhood mortality in Nigeria., 2012, http://www.ppaa.princeton.edu/paper/91346.
[54] Nzewi E. Malevolent ogbanje: recumbent reincarnation or sickle cell disease., 2001, http://www.sciencediret.com/source/article/pii/50277953600002458.
[55] Agara AJ, Makanjuola AB, Morakinyo O. Management of perceived mental health problems by spiritual healers: a Nigerian study. Afr J Psychiatry 2008;11:113–8.
[56] Andrews KM, Brouillette DB, Brouillette RT. Mortality, infant. In: Encyclopedia of infant and early childhood development. Elsevier; 2008. p. 343–59.
[57] Nussbaum M. Creating capabilities. The Belknap Press of Harvard University Press; 2011, ISBN:978-0-674-05054-9.
[58] Jorgenson AK. Global inequality, water pollution, and infant mortality. Soc Sci J 2004;41(2):279–88.
[59] Gortmaker SL, Wise PH. The first injustice: socioeconomic disparities, health services technology, and infant mortality. Annu Rev Sociol 1997;23:147–70.
[60] The State of the World's Children. Child survival. UNICEF; 2008. http://www.unicef.org>state-worlds-children-2.

[61] Russo LX, Scott A, Sivey P, Dias J. Primary care physicians and infant mortality: evidence from Brazil. PLoS ONE 2019;14(5), e0217614.

[62] CDC. Commit to healthy choices to help prevent birth defects | CDC. Centers for Disease Control and Prevention; 2020. Retrieved 2020-07-30.

[63] Atta CA, Fiest KM, Frolkis AD, Jette N, Pringsheim T, St Germaine-Smith C, et al. Global birth prevalence of spina bifida by folic acid fortification status: a systematic review and meta-analysis. Am J Public Health January 2016;106(1):e24–34. https://doi.org/10.2105/AJPH.2015.302902. PMC 4695937. PMID 26562127.

[64] Ramakrishnan U, Grant F, Goldenberg T, Zongrone A, Martorell R. Effect of women's nutrition before and during early pregnancy on maternal and infant outcomes: a systematic review. Paediatr Perinat Epidemiol July 2012;26(Suppl 1):285–301. https://doi.org/10.1111/j.1365-3016.2012.01281.x. PMID 22742616.

[65] Flak AL, Su S, Bertrand J, Denny CH, Kesmodel US, Cogswell ME. The association of mild, moderate, and binge prenatal alcohol exposure and child neuropsychological outcomes: a meta-analysis. Alcohol Clin Exp Res January 2014;38(1):214–26. https://doi.org/10.1111/acer.12214. PMID 23905882.

[66] Banderali G, Martelli A, Landi M, Moretti F, Betti F, Radaelli G, et al. Short and long term health effects of parental tobacco smoking during pregnancy and lactation: a descriptive review. J Transl Med October 2015;13:327.

[67] Anon. Breastfeeding and the use of human milk. Pediatrics March 2012;129(3):e827–41. https://doi.org/10.1542/peds.2011-3552.

[68] Center of Disease Control and Development. Child development. CDC.gov; 2021. http://www.cdc.gov>ncbdd>childdevelopment.

[69] Shapiro D., Tenikue M. (2017-09-13). "Women's education, infant and child mortality, and fertility decline in rural and urban sub-Saharan Africa". Demographic Research 37: 669–708.

[70] Mathews TJ, MacDorman MF. Infant mortality statistics from the 2009 period linked birth/infant death data set (PDF). National Vital Statistics Reports January 2013;61(8):1–27. PMID 24979974.

[71] Anon. Austria infant mortality rate; 1950-2021. http://www.macrotrends.net>countries>AUT>int.

[72] Anon. German infant mortality; 1950-2021. http://www.macrotend.net>countries>DEU>infan.

[73] Anon. France infant mortality rate; 1950-2021. http://www.macrotrends.net>countries>FRA>infan-m.

[74] O'Nell A. Infant mortality rate in Czech Republic; 2019. http://ec.europa.eu>state>docs>chp cs English.

[75] Kail RV, Cavanaugh JC. Human development: a lifespan view. 5th ed. Cengage Learning; 2010. p. 296, ISBN:978-0-495-60037-4.

[76] Schuiling. Women's gynecologic health. Jones & Bartlett Learning; 2016. p. 22, ISBN:978-1-284-12501-6.

[77] Phillips DC. Encyclopedia of educational theory and philosophy. Sage Publications; 2014. p. 18–9, ISBN:978-1-4833-6475-9.

[78] UNICEF. Two third of the world's school-age children have no internet; 2020. http://unicef.org>press-release>two-thir.

[79] Machel G. The impact of armed conflict on children: report of the expert of the secretary general of the United Nations. New York: United Nations; 1996. Available from: http://www.unicef.org/graca/a51-306_en.pdf.

[80] Pearn J. Children and war. J Paediatr Child Health 2003;39:166–72.
[81] US Fund for UNICEF. Landmines pose the greatest risk for children. http://www.unicefusa.org/site/apps/nl/content2.asp?c=duLRI8O0H&b=279482&ct=307827.
[82] ICPD25. Sexual and reproductive health and rights: an essential element of universal health coverage. United Nations Publication Fund; 2019. http://www.untpa.org>default>files>pub-pdf.
[83] Arcury TA, Gesler WM, Preisser JS, Sherman J, Spencer J, Perin J. The effects of geography and spatial behavior on health care utilisation among the residents of a rural region. Health Serv Res 2005;40(1):135–55.
[84] Bernard S, Perry H. Loneliness and social isolation among older people in North Yorkshire. Stage 2 of project commissioned by North Yorkshire older people's partnership board., 2013, https://www.york.ac.uk/inst/spru/research/pdf/Lone.pdf.
[85] Freeman J, McGowan JE. Risk factors for nosocomial infections. J Infect Dis 1978;138:811–9.
[86] Schneider EL. Infectious diseases in the elderly. Ann Intern Med 1983;98:395–400.
[87] Gardner ID. The effect of aging on susceptibility to infection. Rev Infect Dis 1980;2:801–10.
[88] Pot AM, Briggs AM, Beard JR. World Health Organisation (2017) healthy ageing and the need for a longterm-care system. Federal Ministry of Health; 2017.
[89] Symith A, Trevelyan E. The older population in rural America: 2012-2016; 2019. Report number ACS-41.
[90] Chatterji S. World Health Organisation's (WHO) study on global ageing and adult health (SAGE). In: BMC proceedings, Vol. 7, No. S4. BioMed Central; 2013. p. S1.

CHAPTER

Nutrition issues and maternal health 4

Both for men and women, good food, rich in quality nutrients is essential for a healthy life, and to live with dignity and success. Regular intake of a quality diet keeps the body and mind strong and robust. Good quality foods (Fig. 4.1) prevent malnutrition in all its forms.

However, modern lifestyles and use of processed foods bring many unwanted health problems. People commonly prefer to use high-energy foods, such as fats, free sugars, and salt/sodium, instead of fresh fruits, vegetables (Fig. 4.2), and others high in dietary fiber, such as wholegrains.

Food habits are not common to all people, but vary according to age, gender, lifestyle, nature of labor, and geographic location. Healthy diets throughout life promote healthy pregnancy outcomes, support normal growth and development, and reduce the risk of disease contamination.

Women play a significant role in the availability, access to, sustainability, and production of food at a rural level to strengthen food security. As food producers, rural women dedicate their time, incomes and decision-making to maintain food and nutritional security of their households, communities, and nations. In most of the rural areas of developing countries, women are involved in crop production and sale of fruits and vegetables grown on plots they manage. However, due to gender disparities, they denied the opportunity to hold land rights independently of their husbands or male relatives. In many developing countries and least developed countries, statutory law does not allow women's independent rights to be the owner of properly and lands, although they are frequently the major household food producers. There are usually customary provisions for indirect access to land in terms of use rights as their status as wives, mothers, sisters, or daughters.

The contribution of women to rural agriculture development, food security, and rural community microeconomics is an unchallenged issue. They are deeply involved in many vital roles, as farmers, wage laborers, and small-scale entrepreneurs, as well as caretakers of children and the elderly. Rural women are capable of managing households and communities during poverty crises, but, they are persistently being held back by gender inequalities that limit their access to taking part in leadership for community development.

The UN Food and Agriculture Organization (FAO), in collaboration with the WHO, works to reduce gender inequalities in the agriculture sector to minimize hunger, poverty, and injustice in the world. The FAO has been making efforts to

Healthcare Strategies and Planning for Social Inclusion and Development. https://doi.org/10.1016/B978-0-323-90447-6.00003-5

FIG. 4.1

Quality food for good health.

FIG. 4.2

Different types of fresh vegetables.

provide sufficient information and technical assistance to policymakers on formulating guidelines on gender equality, food security, agricultural intensification programs, agricultural extension schemes, technology and infrastructure innovations, agriculture-related investment, and trade and marketing regulations. Many International Funds for Agricultural Development (IFAD)-supported activities working with poorer households recognize and address their core need to improve household food and nutrition security by promoting smallholder production and improving postharvest handling.

4.1 Biological significance of quality food

Life is not sustainable without food. To be healthy we should have balanced, nutrient-rich food in our day-to-day activities. We have to ensure that the food we take in is safe and rich in all the nutrients and essential elements for our body (Table 4.1).

It is advisable to take seasonal fruits and vegetables as natural food supplements along with our primary food. Food fulfills our need to stay alive, be active, move, and work. It also fulfills the biological need of cell division, tissue formation, and overall growth and development of body organs. In addition to supporting the overall function of the body, food protects our body from communicable and noncommunicable diseases.

4.1.1 Significance of quality food

The practice of keeping food safe and ensuring its quality (Fig. 4.3) begins with production and ends with the consumer.

The Food and Agriculture Organization (FAO) is the only international body overseeing all aspects of food supply chain management and has a unique vision for food safety. The World Health Organization and FAO jointly cover ranges of issues to support global food safety and protect consumer health. The WHO plays a significant role in maintenance of strong relationships with the public health sector, whereas the FAO generally addresses food safety issues, including food supply chain management.

4.1.2 Food safety and security

Generally, food safety and security refers to disease-free food (serials, vegetables, and fruits), food free from additives and other chemicals and biological toxins, and food adulteration (Fig. 4.4).

Whereas food quality is comprised of consumers' attitudes, which collectively influence them to put a higher value on, mainly, calories and nutrient density.

Calories present in food are a major source for growth and development of the body. A calorie is actually a unit of energy, and they are found in the foods and drinks that we consume. They are the real source of energy to our body for our day-to-day lives and fuel for high-labor activities. Mainly, calories are present in carbohydrates, fats, and proteins.

Table 4.1 Food vales of some important fruits.

Fruit	Serving size	Calories	Carbs (g)	Proteins (g)	Fibers (g)	Fat (g)	Sodium (mg)
Apple	1 medium size	80	22	0	5	0	0
Peaches	-do-	40	10	0.6	1.5	0	0
Nectarines	-do-	70	16	1	3	1	0
Plums	-do-	36	86	0.52	1.0	0.41	0
Asian pears	-do-	59	13	0.9	4	0.1	0
Strawberries	8 medium size	70	17	18	3	0.5	0
Raspberries	10	60	2.3	0.2	1.2	0.1	0.2
Blueberries	1 cuo	23	21	1.1	3.5	0.5	1
Pumpkin	Cup pulp juice	49	12	2	3	0	0

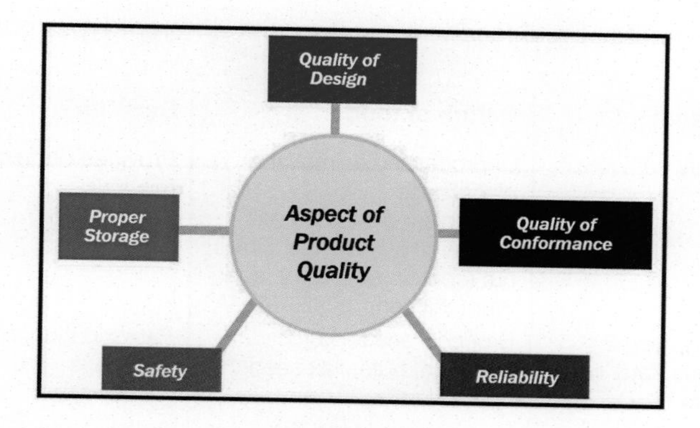

FIG. 4.3

Different aspects of food quality control.

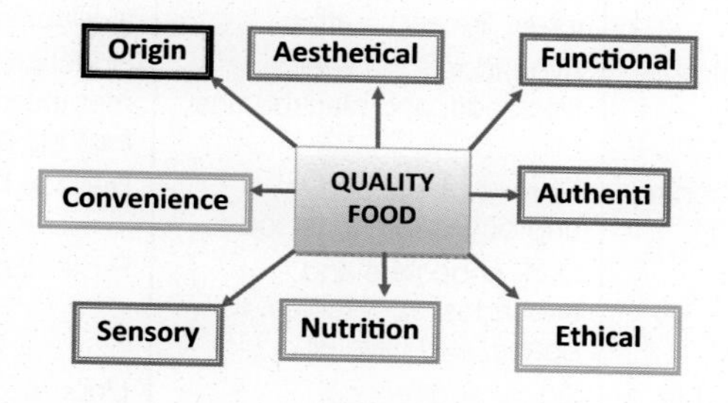

FIG. 4.4

Characteristics of real food.

Nutrient density refers to the quantity of beneficial micronutrients (vitamins and minerals) in food in proportion to its calories. Nutrient-dense foods are typically low in calories but rich in their amount of micronutrients.

The sustainability of healthy and durable life depends on access to adequate quality food. Globally, about one in 10 people fall ill after eating contaminated food each year, resulting in 420,000 deaths and the loss of 33 million healthy life/years.

Food safety, security, and quality nutrients are interlinked factors. Unsafe food is responsible for creating a vicious cycle of disease and malnutrition, especially affecting infants and elderly persons. In addition, food safety supports tourism, national economies, and trade, and helps in sustainable development. Consequently, the WHO aims to enhance at a global- and country-level the capacity to prevent, detect, and respond to public health threats associated with unsafe food.

4.1.3 Benefits of high-quality foods and regulation

Food quality refers to the characteristics of food acceptable to consumers' requirements (Table 4.2) [1–3].

Table 4.2 Benefits of real food to overcome with different types of health problems.

No.	Requirement	Nature of food	Benefits
1	As body nutrients	Unprocessed animal and plant foods Examples: gram, red bell peppers/kiwi, orange, Brazil nut (contain selenium), egg, liver (chicken)	Provides vitamins and minerals
2	Obesity, type 2 diabetes, fatty liver disease	Broccoli, sea food, pumpkin, nuts, flax seeds/beans, lentils	Fat-oxidation, sugar control
3	For the heart	Unprocessed food like cabbage, tomato, sweet corn	Antioxidants Protect from toxic substances
4	Better environment	Sustainable agriculture, stop forest fire and cutting	For health improvement, availability of good food
5	Stomach control and digestive control	Avocados, chia seeds, flaxseeds, and blackberries	Boosting digestive function, metabolic health, and feelings of fullness
6	Blood sugar control	Fibrous plants and unprocessed animal foods	Reduce blood sugar
7	Skin care	Dark chocolate and avocados, vegetables, fish, beans, and olive oil	Protect skin from wrinkling
8	Blood triglyceride levels	Sugar and refined carbohydrates (carbs)	Unprocessed foods like fatty fish, lean meats, vegetables, and nuts have been shown to significantly reduce triglyceride
9	To provide variety	Diverse foods: all type of vegetables, a variety of meat, fish, dairy, vegetables, fruits, nuts, legumes, wholegrains, and seeds	Growth and development
10	Real food	Unprocessed. free of chemical additives, rich in nutrients, brown rice	Better growth and development
11	High in healthy fat	Olive oil, coconut oil	Promotes heart health
12	Reduced disease risk	Mediterranean diet whole, unprocessed foods	Reduce your risk for heart disease, diabetes, and metabolic syndrome
13	Antioxidants-rich food	Vegetables, fruits, nuts, wholegrains, and legumes. Fresh, unprocessed animal foods also contain antioxidants—though in much lower levels	Protect from singlet oxygen damage, protect against eye diseases like cataracts and macular degeneration

Table 4.2 Benefits of real food to overcome with different types of health problems—cont'd

No.	Requirement	Nature of food	Benefits
14	Good for gut	Real food	Gut micro biomass, which refers to the microorganisms that live in your digestive tract
15	To prevent overeating	Highly processed food	Increase body weight
16	To promote dental health	Real food	Save from dental decay by feeding the plaque-causing bacteria that live in your mouth
17	Sugar cravings	Stop excess sugar intake	Reduce overall growth pattern
18	Sets a good example	Real food	Stay healthy
19	Get the focus off dieting	Good nutrition is about much more than losing weight	Focusing on eating balanced meals rich in fruits and vegetable instead of dieting
20	Help support local farmers	Supports the people who grow food in your community	One can get fresh food
21	Delicious	Real food tastes delicious	For healthy life

Generally, the quality of food is assessed by external features (size, shape, color, gloss, and consistency), texture, flavor (federal grade standard), and internal condition (possibility of contamination and nutrient value). In general, food quality is controlled at the production or manufacturing stage on the basis of consumers' requirements [4].

In developed countries food standards are regulated, for example, in the United States, food quality is enforced by the Food Safety Act 1990. However, food safety regulation is mainly dependent on local or state federal government of a nation. The US Food and Drug Administration (FDA) is responsible for ensuring food products are safe, sanitary, nutritious, and properly balanced. The primary statutes governing the FDA's activities are the Federal Food, Drug, and Cosmetic Act (FFDCA) and the public Health Services Act [5]. State, local, and county public health and agriculture departments play a major role in helping the FDA carry out these responsibilities by training and giving guidelines and financial support.

Production and manufacturing processes of food are mainly based on consumers' requirements. Generally, foods are sensitive to temperature and vulnerable to foreign pathogenic microbes. Proper preservation of food is a highly responsible task. Consumers have a right to expect that the foods they purchase and consume will be safe and high quality. For this, the food ingredient quality, manufacturing standards, and preservation method all should follow the standards set by regulatory acts of the nation concerned or an internationally recognized regulatory agency like the FDA or ISI. Besides ingredient quality, there are also sanitation requirements to ensure food quality from a hygienic point of view. In order to ensure correcting ingredient and nutritional information, the food should be labeled with traceability, name of the product, the manufacturer name and address, net weight, serving size, list of ingredients, and nutrition information per serving.

4.2 Health and nutrition

4.2.1 Health

The Constitution of the World Health Organization, which came into existence on April 7, 1948, defines health as "A state of complete physical, mental and social well-being." But, at a later stage, the writers of the Constitution are clearly aware of the tendency of states to define health only according to the presence or absence of diseases: the definition of health being given as "A state of complete physical, mental and social well-being and not merely the absence of disease or infirmity" [6].

Meanwhile, with the passage of about seven decades, people have begun to reshape the definition of health, as proposed in 1948. Today, three types of definition of health are mentioned. The first is that health is the absence of any disease or impairment. The second is that health is a state that allows the individual to adjust to all demands of daily life including the absence of disease and impairment. The third definition refers to health as a state of balance—an equilibrium that an individual has within himself or herself under the surrounding social and physical environment.

4.2.2 Nutrition

The most widely accepted definition of nutrition, as defined by Robinson, is "it is the science of foods, nutrients and other substances, there in, their action, interaction and balance in relation to health and disease, the process by which the organism ingests, digests, absorbs and utilizes and disposes off their end products."

Nutrition is a functional aspect of the biological process by which a living entity uses food to support life. The overall processes of nutrition include ingestion, absorption, assimilation, biosynthesis, catabolism, and excretion. Real food is the best nutrient for good health (Fig. 4.4). It is primarily defined as being:

- Unprocessed
- Free of chemical additives
- Rich in nutrients

However, currently, due to heavy workloads and other priorities, people prefer to eat highly processed foods and ready-to-eat meals. While processed foods are

FIG. 4.5

Natural vegetables can be eaten without processing.

convenient, they can also potentially affect your health. So following a diet based on real food may be one of the most important things you can do to help maintain good health.

Real foods are otherwise known as natural food (Fig. 4.5). However, the FDA has not developed any rules or regulations on the defining features of what qualifies a product as "natural." Generally, people perceive natural food as "nothing artificial or synthetic (including colors and edibility nature)."

The Mediterranean diet is also based on natural food, which can be eaten without processing. It is high in vegetables, fruits, legumes, nuts, beans, cereal, grain, fish, and unsaturated fats such as olive oil. It includes a low intake of meat and dairy foods. A Mediterranean diet incorporates the traditional healthy living habits of people from countries bordering the Mediterranean Sea, including France, Greece, Italy, and Spain. The Mediterranean diet varies by country and region, so it has a range of definitions. The Mediterranean diet has been linked with good health, including a healthier heart.

Reasons for eating real food are tabulated in Table 4.2. Proper nutrition is critical for growth and development, especially for improvement of infant, child, and maternal health, and promotes a strong immune system, safe pregnancy and childbirth, and lower risk of noncommunicable diseases (Table 4.3). Malnutrition, in every form, is detrimental to human health. Currently, we face a double burden of malnutrition that includes both under-nutrition and overweight, especially in low- and middle-income countries.

A regular intake of healthy and balanced food from the beginning of life helps to prevent malnutrition in all its forms as well as different types of noncommunicable diseases (NCDs). However, greater availability of processed foods, rapid urbanization, and changing lifestyles has led to a shift in dietary patterns. Now, the preferred diet of many people is high energy fatty food with free sugar and large

Table 4.3 Nutrient value of different types of fruits for health benefits.

	Type of food	Sugar	Vitamin	Protein/fiber/ antioxidant
1	Lemons (and limes)	1–2 g/lemon	High vitamin C	
2	Raspberries	Teaspoon of sugar per cup	Vitamin C	Rich in fiber. Water-soluble nutrient essential for immune function and iron
3	Strawberries	7 g/cup of juice	High vitamin C	
4	Blackberries	Seven grams Trusted source of sugar per cup		Antioxidants as well as fiber
5	Kiwis	Six grams Trusted source per kiwi	Vitamin C	
6	Grapefruit		Vitamin C	Rich source of antioxidants
7	Avocado	One gram trusted source of sugar/fruit		Rich in healthy fats
8	Watermelon	Cup of diced up watermelon has under 10 g Trusted Source		Great source of iron
9	Cantaloupe	Delicious melon contains less than 13 g	High vitamin A	
10	Oranges	12 g per fruit	Vitamin C	
11	Peaches	Less than 13 g of sugar in a medium-sized fruit		
12	Pomegranates	One cup containing 24 g of sugar and 144 cal	Good source of vitamin C, vitamin K, and potassium	Pomegranates are rich in antioxidants and flavonoids One cup of fruit juice (174 g) contains [2]: Fiber: 7 g. Protein: 3 g. Vitamin C: 30%
13	Banana medium size	15 g sugar		0 g fat, 1 g protein, 28 g carbohydrate, 3 g fiber, and 450 mg potassium

quantities of salt and sodium; whereas, people do not habitually eat enough fruit, vegetables, and other sources of dietary fiber such as wholegrains. Generally, the habit of eating healthy food and a balanced diet depends on the individual, on the basis of age, gender, lifestyle, and daily nature of physical activity. Furthermore, the cultural context, locally available foods, and dietary customs also influence the food habits of an individual. The following are a few important tips for nutrient-rich food for adults and infants.

4.2.2.1 Adults

To maintain a healthy diet for an adult:

- The day-to-day diet should commonly include fruit, vegetables, legumes (e.g., lentils and beans), nuts, and wholegrains (e.g., unprocessed maize, millet, oats, wheat, or brown rice).
- There should be a proper balance between energy intake and energy expenditure. Intake of fatty food should not exceed 30% of energy intake [7–9]. The intake of saturated fat should be limited to 10% of total energy intake, and intake of trans-fats should not be more than 1% of total energy intake. It is also necessary to avoid commercialized trans-fats [10–12]. Total carbohydrate per day should not exceed 10%, which is equivalent to 2000 cal per day. Reduction in free sugars intake to less than 5% of total energy intake would provide additional health benefits [13].
- Overeating of high sugar (free sugar) content diet may lead to type 2 diabetes.
- There should be restriction in consumption of salt (equivalent to sodium intake of less than 2 g per day) to less than 5 g per day. This helps in prevention of hypertension, and reduces the risk of heart disease and stroke in the adult population [14].

WHO member states have agreed to reduce the global population's intake of salt by 30% by 2025. They have also agreed to halt the rise in diabetes and obesity in adults and adolescents as well as childhood overweight by 2025 [15,16].

4.2.2.2 Infants and young children

Advice on a healthy diet for infants and children is similar to that for adults, but the following elements are also important.

The target of Sustainable Development Goal-3 is to bring down the neonatal mortality to at least as low as 12 per 1000 live births by the end of 2030. Even with the improvement of medical sciences and technology, under nutrition is associated with 2.7 million child deaths annually or 45% of all child deaths. The main reason for this is lack of proper nutrition care for infants and young children. Infant and young child feeding is a key practice in women to improve child survival. The first 2 years of life are particularly important, as optimal nutrition during this period lowers morbidity and mortality, reduces the possibility of chronic disease, and fosters overall better development. About 820,000 deaths/year of children under 5 years could be saved by optimal breast feeding. Both the WHO and UNICEF recommend:

- Early initiation of breastfeeding, within 1 h of birth;
- Exclusive breastfeeding for the first 6 months of life; and
- Introduction of nutritionally adequate and safe complementary (solid) foods at 6 months, together with continued breastfeeding up to 2 years of age or beyond.

However, many infants and children do not receive optimal feeding. On the basis of both WHO and UNICEF reports, during 2015–20 about 44% of infants aged 0–6 months worldwide did not receive breastfeeding [17]. Recommendations for HIV-infected mothers have been made to supply antiretroviral drugs to allow their infants to be exclusively breastfed until they are 6 months old and continue breastfeeding until at least 12 months of age, providing a significantly reduced risk of HIV transmission.

Exclusive breastfeeding for 6 months has many benefits for the infant and mother. The practice of breastfeeding protects the infant against gastrointestinal infections, mostly observed in rural areas of developing countries. Early initiation of breastfeeding, within 1 h of birth, protects the newborn from acquiring infections and reduces newborn mortality. Breast-milk is also an important source of energy and nutrients to infants, even up to 6–23 months. Children and adolescents who were breastfed as babies are less likely to be overweight. Longer duration of breastfeeding also helps the health and well-being of the mother: it reduces the risk of ovarian and breast cancer and helps space pregnancies.

After the completion of 6 months, a baby needs more energy and nutrients, due to increase in body growth and development. So, at this stage, the mother's milk alone will not be sufficient. Complementary food is necessary. So, it is necessary to feed baby food like boiled cereals, barely, and pulp of different types of fruits, but in small amounts without any force. Gradually a child will be habituated with supplementary food rich in nutrients and essential elements required for growth and development. Guiding principles for appropriate complementary feeding are:

- Feed breast milk on demand of baby;
- Develop the habit of taking a variety of complementary food, gently;
- Practice hygiene and balanced diet with the advice of healthcare providers;
- Slowly increase the amount of complementary food with the increase of age and the advice of a doctor;
- Use fortified complementary foods or vitamin- mineral supplements as needed;
- During illness, increase fluid food intake, rich in nutrients.

4.3 Requirements of nutrition for rural women

The overall well-being of the family depends on the leadership of healthy women. The issues of maternal morbidity and mortality not only affect the mother but also the entire family. Besides taking care of home responsibility, women take care of community development, and rural microeconomic and village supply chain management of agriculture products. So, it is obvious that women's health and nutritional status is of significant importance for their own family, as well as the development of a nation. Rural women with poor health cannot shoulder the entire responsibility of family members and healthcare of children.

In general, it is everybody's right is to have access to quality health, and this is especially so for women, as they constitute about half of the world's population and more than one-third of the workforce. In addition, a major responsibility of women is rearing the children of the family. A healthy woman can only eligible to bring social sustainability in terms of jobs, food, energy, water, and sanitation. Inadequate nutrients makes women prone to certain ailments. Poor health status of women leads to lower life expectancy, high rates of morbidity and mortality, lower levels of productivity, and decrease in earning efficiency. Functionally, we can study food focusing on three major issues, as discussed below.

4.3.1 Energetic foods need for high laborious work

Vegetables and fruits grown on family farmland may hold the key to improving diets and nutritional status of rural women, but due to shortage of money to maintain health, they sell their fruits and vegetables. This is mainly due to proximity of villages to cities, and increasing contact with people from other places. So, it is necessary for governments to supply rural people with a variety of cereals (wheat, rice, jowar, maize, and bajra) on the basis of subsidy or supply fertilizers with lower value, in order to reduce the burden of agriculture expenditure during adverse climatic conditions.

Generally, rural women spend more energy on managing fuel and water for regular household work. There are pockets in India where women have to work hard for hours to gather water and firewood. Rural women put a huge amount of labor into the simple task of cooking that is never quantified. A bundle of firewood that a woman carries amounts to between 8 and 10 kg, the weight depending on the size of the family. If the family owns cattle, the woman's load is reduced a bit, as dried cow dung can be used to fire the stove. One of the major problems associated with this enormous physical burden is missing opportunities for other developmental activities. This tiring and prolonged work takes a toll on the women's health. Most often, it is men who migrate to urban areas for work and the women are forced to work hard to attend to all the needs of the family. With the passing of time and problems of aging, make women are helpless in the face of chronic health problems.

4.3.2 Foods for growth and maintenance

Due to poverty, it has been difficult to meet the need for a balanced diet for rural women at the pregnancy stage. In order to overcome this basic health problem, governments must provide a nutrient-rich diet to rural women either, in terms of subsidy or free, based on the financial status of the government.

For rural women, mainly in developing countries, the requirement of nutrition and a balanced diet for female in their entire life cycle is affected by social, cultural, and health service-related factors. Therefore, it is necessary that, instead of focusing only on the prenatal period, an entire life cycle approach to improve maternal nutrition be taken. It is high time that the traditional provision of nutrition services during pregnancy should be extended to before pregnancy. It is also necessary to fix the "minimum package" of nutrition diet supply system in rural primary health services centers. It is well known that an undernourished mother inevitably gives birth to an undernourished baby. Under nutrition is a deficiency of calories or of one or more essential nutrient. Under nutrition may develop because people cannot obtain or prepare food, have a disorder that makes eating or absorbing food difficult, or have a greatly increased need for calories.

A balanced diet should contain:

- Energetic proteins
- Complex carbohydrates
- Healthy types of fat (monounsaturated fat or polyunsaturated fat)
- Vitamins and minerals
- Fiber and fluid

A healthy pregnancy eating pattern contains much the same balance of vitamins, minerals, and nutrients as healthy eating patterns in general. Animal sources like milk and milk products, eggs, meat, fish, etc., are rich sources of protein, Cereals also contains protein, although in smaller amounts.

4.3.3 Food for protection against diseases

Some of the important disease-fighting foods are dark, leafy greens, which include spinach, kale, and dark lettuces. They are rich with vitamins, minerals, beta-carotene, vitamin C, folate, iron, magnesium, carotenoids, phytochemicals, and antioxidants. Some important disease-fighting foods commonly available are as follows.

i) Berries

Commonly, berries are rich in antioxidants, which can reduce the quantity of singlet oxygen in the body. According to a US Department of Agriculture study, blueberries top the list of antioxidants fruits, followed by cranberries, blackberries, raspberries, and strawberries. Dark berries and blackberries lower the level of cholesterol and are good for cardiovascular health. The berries are rich in anthocyanin, which is an antioxidant that neutralizes free radicals in the body. Berries, particularly cranberries, may help ward off urinary tract infections.

ii) Dairy

Dairy foods are rich in calcium, with plenty of proteins, vitamins (including vitamin D), and minerals. Dairy foods are good for fighting osteoporosis. The US Government's 2005 Dietary Guidelines recommend having three daily servings of low-fat dairy products, as well as doing weight exercise, to help keep bones strong. Those who cannot digest dairy food can take other calcium-containing foods include legumes; dark green leafy vegetables such as kale, broccoli, and collards, and calcium-fortified soy products, juices, and grains.

Dairy products also help for weight loses. Low-fat dairy foods are very good for snacks, as they contain both carbohydrates and proteins.

iii) Fatty fish

Fish like salmon and tuna are rich in Omega-3 fatty acids, which help in preventing blood clots associated with heart disease. Eating a diet rich in fatty fish can help reduce the risk of cardiovascular disease.

iv) Dark, leafy greens

Spinach, kale, bok choy (Chinese cabbage), and dark lettuces are rich in vitamins, minerals, β-carotene, vitamin C, folate, iron, magnesium, carotenoids, phytochemicals and antioxidants. Spinach helps prevent type 2 diabetes.

v) Wholegrains

Oatmeal contains soluble fiber, which helps in lowering blood cholesterol level. Generally, while refining wholegrains, the nutritional components are stripped away from the grain surface. They contain folic acid, selenium, and B vitamins, and are important for heart health, weight control, and reducing the risk of diabetes. The fiber content of wholegrain helps the individual to feel "full," which is ultimately helpful in weight loss.

vi) Sweet potatoes

These luscious orange tubers contain a wealth of antioxidants and phytochemicals, including beta-carotene, vitamins C and E, folate, calcium, copper, iron, and potassium. The fiber in sweet potatoes promotes a healthy digestive tract, and the antioxidants play a role in preventing heart disease and cancer.

vii) Tomatoes

The red color of tomatoes is due to the presence of lycopene, an antioxidant that fights cancer. They also deliver an abundance of vitamins A and C, potassium, and phytochemicals. Tomatoes can be enjoyed raw, sliced, cooked, chopped, or diced as part of any meal or snack.

viii) Beans and legumes

Beans and legumes are packed with phytochemicals; fat-free, high-quality protein; folic acid; fiber; iron; magnesium; and small amounts of calcium. Beans are a very good source of inexpensive proteins, otherwise known as poor men's protein.

Regular dietary intake of beans and legumes is good for a healthy life and reduces the risk of certain cancers, lowers blood cholesterol and triglyceride levels, and stabilizes blood sugar.

ix) Nuts

Nuts contain mono- and polyunsaturated fats, which can help lower cholesterol level and prevent heart disease. In addition, nuts are a good source of proteins, fiber, selenium, vitamin E, and vitamin A. A small amount of nuts can boost energy and beat hunger, helping diets stay on track.

x) Eggs

Although egg contains a small amount of fat, they can play a significant role in elevating blood cholesterol; however, eggs are rich in high-quality protein, and are an excellent source of lutein and choline, which are helpful to pregnant women. Eggs are also helpful for eye health and prevent age-related macular degradation, the leading cause of blindness in old people.

4.4 Support for rural women in household management

i) To provide access to promote household work

This is mainly to promote access to a generalized household food ratio through a public distribution system, and also to make provision for supplying supplementary foods under the integrated child development services scheme. Integrated Child Development Services was launched by Government of India in 1975 to improve the health, nutrition and education, simultaneously of children up to the age 6 years. To provide knowledge to improve the local diet, production, and household behaviors through nutrition and health education.

ii) Preventing micronutrient deficiencies and anemia

Rural women face serious threats to health due to inadequate availability of vitamin A (vitamin A deficiency, VAD), iron (iron deficiency anemia, IDA), and iodine (iodine deficiency, IDD). Three major intervention strategies available for the control of micronutrient malnutrition are: (i) supplementation of the specific micronutrients; (ii) fortification of foods with micronutrients; and (iii) horticulture intervention to increase production and educate the women to take food rich in micronutrients regularly. Fortunately, in many developing countries like India, the government has started providing iron, vitamin A and folic acid tablets, and salt iodization through primary health care centers.

iii) Promoting women's access to basic nutrients

To develop the provision of early registration of pregnancy and quality of antenatal checkup, with emphasis on pregnancy weight gain monitoring, screening, and special care of at-risk mothers.

iv) Access to clean water and sanitation

It is of utmost importance to educate rural women on sanitation, and in this connection special attention should be paid to hygiene education related to menstrual hygiene.

v) To prevent early marriage

Early marriage should be discouraged thru a campaign to raise awareness and ensuring girls complete secondary education. It is also important to prevent maternal depletion by delaying first pregnancy and repeated pregnancies through family planning, reproductive health information, incentives, and services. In addition, it is necessary to promote community support systems for women to support decision making, confidence building, skill development, and economic empowerment.

4.5 Significance of health and nutrition in women's life cycles

As compared to urban women, rural women face more intensity in poverty, hunger, gender discrimination, malnutrition, illiteracy, and hard physical work in their life cycle. In addition, the practice of early marriage and repeated and multiple pregnancies make a rural woman's life helpless. Even after taking responsibility of entire household work, female suffer high rate of mortality. Apart from having nutritional deficiency, female suffer from lot of gynecological problems due to inadequate primary healthcare services in the village community. The following are some of the important factors faced by rural women throughout their life cycle.

4.5.1 Gender bias

Gender bias refers to a preference for one gender over another. It is often based on prejudices and stereotypes. For example, the disparity between female pain treatment and male pain treatment has drawn attention. Commonly, women experience more chronic pain as compared to men. Often, the pain being received by women becomes severe and frequent, but they are less likely to receive attention.

Gender bias low sex-ratio is the indication of discrimination against females. It is mainly noticed in rural areas of developing countries like India. Gender bias can occur due to amniocentesis or finding sex before the birth of a baby, followed by abortion if the child is female. Although, abortion is an illegal practice, but still is existing in the higher society of urban areas. This may be, due to unintended pregnancy, and strict implementation of Medical Termination of Pregnancy (MTP) Act, people from the urban area of India go for illegal abortion

In many developing countries, the female are excluded from the right to inherit parental property. During the distribution of resources at the household level, priority is given to males. Women in poor rural communities are found to seek medical assistance for their sons more frequently than their daughters. The sociocultural milieu is such that women are not expected to bother about their own health. Moreover, women are the last in the family to eat after serving all other family members. This in itself is a reflection of the status accorded to women by the society. By virtue of what is considered their "noble practice," women try to provide the best nutrition to the family members at the cost of ignoring their own needs in cases of food deficiency. So, the rural women have inheritance quality of sacrifice the entire life by taking care of financial burden of family by working at home and supporting male workers in various agricultural practices, taking care of health of elderly member of family and children, and other household works.

4.5.2 Adolescent pregnancy

Adolescent pregnancy (AP) is an extremely a social risks, and individual's misfortune. This is mainly due to that early childbearing may result in poor health outcomes for both mother and child, and it also has social consequences like early school dropout which may lead to lesser ability to earn sufficient money to lead good health and family wellbeing. AP is a leading cause of mental health problems and is associated with many health complications [18–22]. About 3.9 million unsafe abortions among girls aged 15–19 years occur each year, contributing to maternal mortality, morbidity, and lasting health problems [23].

Adolescent pregnancies are a chronic global problem occurring in high-, middle-, and low-income countries. Around the world, however, adolescent pregnancies are more likely to occur in marginalized communities, commonly driven by poverty and lack of education and employment opportunities [24].

A wide range of factors are responsible for adolescent pregnancies. In many developing countries, especially in rural communities, girls under the age of 18 are forcefully married and bear children early [25–27]. About 39% of girls in least developed countries get married before 18 years of age and 12% before age 15 [28]. In many places girls choose to become pregnant because they have limited educational and employment prospects. Often, in such societies, motherhood is valued and marriage and childbearing may be the best of the limited options available.

Due to lack of knowledge on family planning, misconceptions on where to obtain contraceptive methods and how to use the [29] are also important factors for adolescent pregnancy. In addition, adolescents face barriers to access contraception including restrictive laws, policies regarding provision of contraceptives based on age or marital status, health worker bias, and adolescents' own inability to access contraceptives due to lack of knowledge, transportation, and financial constraints.

Early-aged married women are prone to high risks of maternal mortality and morbidity due to complications during pregnancy and childbirth. At this early stage, women are little aware of their sexual and reproductive health and rights (SRHR). Early marriage is one of the major factors responsible for adolescent pregnancy, which leads to serious health risks, and may increase the risk of contracting sexually transmitted infections. In many African countries child marriage is closely linked to female genital mutilation/cutting (FGM/C), which is a human rights violation to girls' physical and mental health.

Adolescence, 14–18 years, is a period of active growth. According to National Nutrition Monitoring Bureau data, during this period, on an average, a girl gains 6.8 kg and 5 cm in their weight and height, respectively. When they are forced into marriage and motherhood at an early age, their natural growth is affected, leading to problems like severe anemia.

In their early stages of the Millennium Development Goals, prevention of adolescent pregnancy and related mortality and morbidity problems were ignored due to unknown reasons [30]. During this period, the WHO, in partnership with other agencies, made a survey and developed the WHO's guidelines for preventing early pregnancy and poor reproductive outcomes in adolescents in developing countries [31]. As the world has transitioned to the Sustainable Development Goals era, adolescents have moved to the center of the global health and development agenda [32]. In continuation of the earlier initiatives of the SDGs, the WHO works closely in partnership within and outside the United Nations to contribute to the global efforts to prevent children becoming wives and mothers. The WHO started putting efforts into evidence-based action, and to support the application of the evidence through well-designed and well-executed national and subnational programs. For example, the WHO works closely with UNICEF, UNFPA and UN Women on global programs to accelerate action to end child marriage. Many nongovernmental organizations have been actively involved in preventing adolescent pregnancy in many countries through innovative projects. There are many small but highly dedicated government-led national programs, e.g., in Chile, Ethiopia, and the United Kingdom.

4.5.3 Quality of antenatal care

Antenatal care refers to the maintenance of physical and mental health of pregnant women with the outcome of sustaining a healthy body, which would be able to deliver a healthy child and lead to a successful and comfortable mother with a healthy baby [33]. At the beginning of antenatal care it is essential to ensure the availability of quality medical care facilities, monitor risk factors, and identify any negative health signs or behaviors through all possible standard medical techniques, and a well-trained health workforce [34].

Every pregnancy carries a risk of complications and some pregnancies carry more risk than others [35]. Generally, antenatal care is taken periodically in a regular sequence. These critical time are a first visit at 8–12 weeks gestation, a second visit at 24–36 weeks gestation, a third visit at 32 weeks, and a fourth visit at 36–38 weeks. However, on the basis of possibility of any health risk, the frequency of visit may be increased. The customary approach emphasizes the number of ANC visits (quantity), while the latter approach acknowledges the importance of quality care [36].

Antenatal care is essential for protecting the health of women and their unborn children. By frequent visits of healthcare providers, women can learn about health behaviors during pregnancy, and also about risk-based symptoms and their possible solutions. In addition, by keeping continuous touch with pregnant women's health, they feel emotional and psychological support at this critical time in their lives. Through antenatal care, pregnant women can accesses micronutrient supplementation, treatment for hypertension, and immunization against tetanus. Antenatal care can also provide HIV testing and medications to prevent mother-to-child transmission of HIV. In malaria, an endemic stage health work force provides pregnant women with medications and insecticide-treated mosquito nets to prevent mosquito attack.

According to the WHO's new recommendation, pregnant women should have at least eight contacts with healthcare providers [37]. These sequential contacts should start from the first 12 weeks gestation, with consecutive contacts at 20, 26, 30, 34, 36, 38, and 40 weeks of gestation. These frequent contacts between pregnant women and health workers increase the satisfaction of mothers [38]. However, in many developing countries like Ethiopia, due to lack of sufficient numbers of health workers, poor infrastructure, and administrative weakness at facilities, it has been difficult to fully satisfy this recommendation [39–41].

As quality care is the most important issue, in the Sustainable Development Goals (SDGs) a minimum index for ensuring increase in antenatal care has been fixed [10]. Quality is very difficult to define owing to the complex nature of the concept and is difficult to measure directly [42–44]. However, on the basis of overall practice, quality care should be at an acceptable cost and capacity to satisfy the needs of the client or patient [45]. In this connection quality antenatal care is considered the care taken during the number of attendance given to all types of patient to deliver quality service [46,47].

On the basis of WHO recommendations, antenatal care (ANC) can be categorized into three groups: (i) assessment (consisting of history taking, physical examination, and laboratory tests); (ii) health promotion (quality nutrients intake, planning the birth, detailed information on pregnancy, subsequent contraception and breastfeeding, and immunization); and (iii) care provision (tetanus toxic immunization, psychosocial support, and recordkeeping) [48]. In spite of a variety of strategies about the content of ANC in different countries, the WHO recommends a core set of services, which include blood pressure measurement, tetanus toxic vaccination, urine testing, iron tablet supplementation, body weight measurement, and counseling about danger signs.

However, due to poor accessibility, poor provider-client interaction, lack of facility resources like inadequate equipment and drugs, lack of qualified professionals, and other administrative issues [49–53], a large proportion of women, especially from rural areas of developing countries, do not receive the minimum four visits in connection with antenatal care [54]. This may also be linked to individual socio-economic and reproductive characteristics, like educational attainment, household, wealth, religion, parity, age, and marital status [55–57]. In this connection, a substantial quantity of facts-based information is available from studies on ANC issues of various developing countries like Ethiopia, the Hadiya Zone, and the Southern Nations and Nationalities of Peoples' Region (SNNPR) [58–63].

Antenatal care in the long run is helpful in reducing maternal morbidity and mortality, as it can identify risk factors and accordingly provide for remedial action. The World Health Organization (WHO) recommends for a minimum of four antenatal care to be given to every pregnant women to reduce perinatal mortality and improve women's experience of care. Certain conditions, like age below 18 years or above 35 years, height below 145 cm, first pregnancy or pregnancy after more than four children, anemia, blood pressure, poor obstetrical history, abortions, stillbirth, multiple pregnancy, etc., often contribute to high maternal morbidity and maternal mortality rates.

In spite of so much provision of antenatal services, in many developing countries like Nigeria, as compared to urban areas, high rates of mortality have been noticed [64–67] in rural areas where, for the most part, antenatal services reach rural expectant mothers only toward the second half of the pregnancy. The Federal Government of Nigeria has adopted primary health care (PHC) as a policy to achieve universal health coverage for citizens and to ensure that women, especially in rural areas, gain access to evidence-based skilled pregnancy care for the prevention of maternal morbidity and mortality [68,69].

4.5.4 Health literacy

Health literacy means the skills of reading, writing, and numeracy in the health domain. However, presently, health literacy refers to a multidimensional concept related to understanding and explaining health for a sustainable normal life.

In 1974, the term health literacy was first used in the proceedings of a health education conference explaining health education as a social policy issue [70]. Subsequently, people started defining health literacy in various ways. In due time, a better understanding of people's health literacy has developed and plays crucial role in adapting health-related information and services and designing successful health education programs.

In widely acceptable terminology, health literacy is the ability to read, understand, and use healthcare information in order to make appropriate health decisions and follow instruction for treatment. However, there are many ways we can define health literacy [71]. Health literacy can be seen in the context of health request demands (e.g., healthcare, media, and internet or fitness facility) or the healthcare skill [72].

Health literacy is a primary contributing factor in health disparities, which has been an increasing concern within the health profession. About a decade back, the National Assessment of Adult Literacy (NAA) conducted a program and found that 36% of participants scored as either "basic" or "below" in term of their health literacy.

The U.S. Department of Health and Human Services (HHS) defines health literacy as "the degree to which individuals have the capacity to obtain, process, and understand basic health information needed to make appropriate health decisions" [73,74]. Health literacy can enable people to read and comprehend essential health-related materials. The practice of health literacy is helpful in understanding their own health and having the capacity to take responsibility for their family's health. About 36% of adults in the United States have low health literacy, with disproportional rates found among lower-income Americans eligible for Medicaid. Individuals with low health literacy experience greater health care use and costs compared to those with proficient health literacy. Over a decade, low health literacy was estimated to cost about $236 billion every year [75].

Generally, low health literacy cannot be seen, but while dealing with patients it can be realized through their conversion and expression. So, healthcare providers should assume that every individual may have difficulty in understanding healthcare information, and use universal precautions to reduce the complexity of their verbal and print communications to reach more patients effectively.

The racial, ethnic, religious, social, and/or linguistic communities also have influence on expressing and understanding health related problems. Therefore, the personal and collective values of an individual should be considered while dealing with patients. The cultural background of an individual patient may reflect their feelings and the way they define health-related problems. It should be the responsibility of a health worker how to perceive the expression of these patients and symptoms to take immediate necessary action. Even individuals with sufficient reading, writing, and numeracy skills can have trouble accessing health services, communicating with healthcare providers, and pursuing effective self-management. The main reason for health disparity is due to cultural mismatches associated with low socioeconomic levels. National estimates reveal that minority populations tend to have greater rates of low health literacy. Out of several factors responsible for health literacy, the following factors have been shown to strongly increase this risk: age (especially patients

65 years and older), limited English language proficiency or English as a second language, less education, and lower socioeconomic status. Patients with low health literacy understand less about their medical conditions and treatment and overall report worse health status.

The health behaviors of persons with low health literacy can be improved by various interventions, such as simplifying information and illustrations, avoiding complicacy, and using "teach-back" methods. The teach-back method, which is also called the "show-me" method, is a communication conformation method used by healthcare providers to conform whether a patient (or care takers) understands what is being explained to them. If a patient understands, they are able to "teach-back" the information accurately.

4.6 Health indicators

Health indicator refers to quantifiable characteristics of a population expressed in terms of the health of that population. Generally, statistical survey analysis on various aspects of health is made to understand which biological indicator is suitable for expressing population health dynamic. Various signals, related to biological activities, like death rate, life expectancy, infant mortality, maternal mortality rate, and proportional mortality rate, can be used as health indicators of population dynamics.

4.6.1 Life expectancy

Life expectancy refers to the average number of years that a person is expected to live. Life expectancy is a hypothetical measure. Life expectancy is affected by many factors such as socioeconomic status, including employment, income, education, and economic well-being; the quality of the health system and the ability of people to access it; health behaviors such as tobacco and excessive alcohol consumption, poor nutrition and lack of exercise; social factors; genetic factors; and environmental factors including pollution, overcrowded housing, lack of clean drinking water, and adequate sanitation.

At present, female human life expectancy is greater than that of males, despite females having higher morbidity rates. There are many potential reasons for this. In practice, males consume more tobacco, alcohol, and drugs than women in most societies, and are more likely to die from many associated diseases such as lung cancer, tuberculosis, and cirrhosis of the liver. Besides, men are also more likely to die from injuries, whether unintentional (such as occupational, war, or car accidents) or intentional (suicide). In 2018, the average life expectancy of women in Japan was approximately 87.3 years, whereas the life expectancy of men in Japan reached almost 81.3 years. The average life expectancy of both men and women in Japan has indicated a continuous growth since 2011.

Japan has the highest proportion of senior citizens worldwide. About 28% of the country's population is aged 65 years and over. Both the growing average life expectancy as well as declining fertility rates has led to this demographic shift. Currently, the retirement age of Japanese businesses is 60 years, however, 50% of senior citizens aged 70 years and over are reported to be either working or involved in volunteer and community activities.

4.6.2 Sex ratio in populations

Sex ratio is a sensitive indicator of the status of women in any society, especially in many developing countries like India. Since 2001, a continuous decline in sex ratio in India has been noticed and developed great concern in the society.

India's sex ratio, or the number of females per 1000 males, declined to 896 in 2015–17 from 898 in 2014–16, acceding to government survey. This may be due to sociocultural factors associated with declining sex ratios, which vary significantly in different parts of the country. However, in developed countries like United States in 2020 were 162,826,299 or 162.83 million males and 166, 238, or 166.24 million females. The percentage of the female population is 50.52% compared to 40.48% male population (UN, world population prospects 2019). In 2020, under 39 years and in the 50–54 years group, the male population is greater than the female. However, in the older age group, 60–64 there are five fewer men per 100 women. The sex ratio at birth is 105 boys per 100 girls.

In Europe the overall sex ration male vs female in different age groups is tabulated as in Table 4.4.

4.6.3 Low weight and height

A majority of women belonging to the lower socioeconomic group are undernourished. The indicators stunting, wasting, overweight and underweight are used to measure nutritional imbalance. Child growth is internationally recognized as an important indicator of nutritional status and health in a population. This measure can therefore be interpreted as an indication of poor environmental conditions or long-term restriction of a child's growth potential. Stunting, wasting, and overweight in children aged under 5 years are included as primary outcome indicators in the core

Table 4.4 Sex ratio in Europe based on different age groups.

Age group	Ratio (male:female)[a]
At birth	1.06
0–14 years	1.05
15–54 years	1.02
55–64	0.95
65 years and above	0.75
Total population	0.96

[a] *CIA World Factbook—This page was last updated on January 20, 2018.*

set of indicators for the Global Nutrition Monitoring Framework to monitor progress toward reaching Global Nutrition Targets 1, 4, and 6. These three indicators are also included in the WHO's Global reference list of 100 core health indicators.

In May 2012, the Sixty-Fifth World Health Assembly endorsed the Comprehensive Implementation Plan (2012–2025) on Maternal, Infant, and Young Child Nutrition, which included global targets on six nutrition indicators: stunting, anaemia, low birthweight, overweight, breastfeeding, and wasting.

4.6.4 Anemia

Anemia is a serious global public health issue by which young children and pregnant women are affected. As reported by the WHO about 42% of children under 5 years of age and 40% of pregnant women worldwide are anemic. Anemia can cause a range of symptoms including fatigue, weakness, dizziness, and drowsiness. In anemia, the number of red blood cells or the hemoglobin concentration is seen to be lower as compared to normal healthy children or pregnant women of that age group. Hemoglobin carries oxygen to body tissues. The most common causes of anemia include nutritional deficiencies, particularly iron deficiency, but also deficiencies in foliate, and vitamins B12 and A. Infectious diseases, such as malaria, tuberculosis, HIV, and parasitic infection are also responsible for anemia. Women who are menstruating or pregnant and people with chronic medical conditions are most at risk from this disease. The risk of anemia increases as people grow older (Fig. 4.6).

Iron deficiency (ID) is estimated to be the most common cause of anemia worldwide and is particularly prevalent in developing nations in Africa and Asia. Other nutritional deficiencies certainly play a role in the occurrence of anemia, but the global prevalence data for these deficiencies are limited.

4.6.5 Maternal mortality

Maternal mortality (MM) has been a great concern over the last 2 decades all over the world (Fig. 4.7).

It is expressed in terms of the ratio obtain by dividing the number of maternal deaths by total recorded live births in the same period and multiplying by 100,000. A number of international conferences have given priority to finding facts and figures on various causes of maternal mortality and possible remedial measures. Some of the top international conferences that have actively discussed the increasing scenario of maternal mortality rate include: World Summit for Children, held in 1990; the 1995 World conference for Women; and the 1994 International Conference on Population and Development.

4.6.5.1 Action plan

Initiation of an action plan on maternal mortality rate at the global level was conducted through the fifth United Nations (UN) Millennium Development Goals (MDGs) in 2000. The target of MDG 5 was to minimized maternal mortality rate by

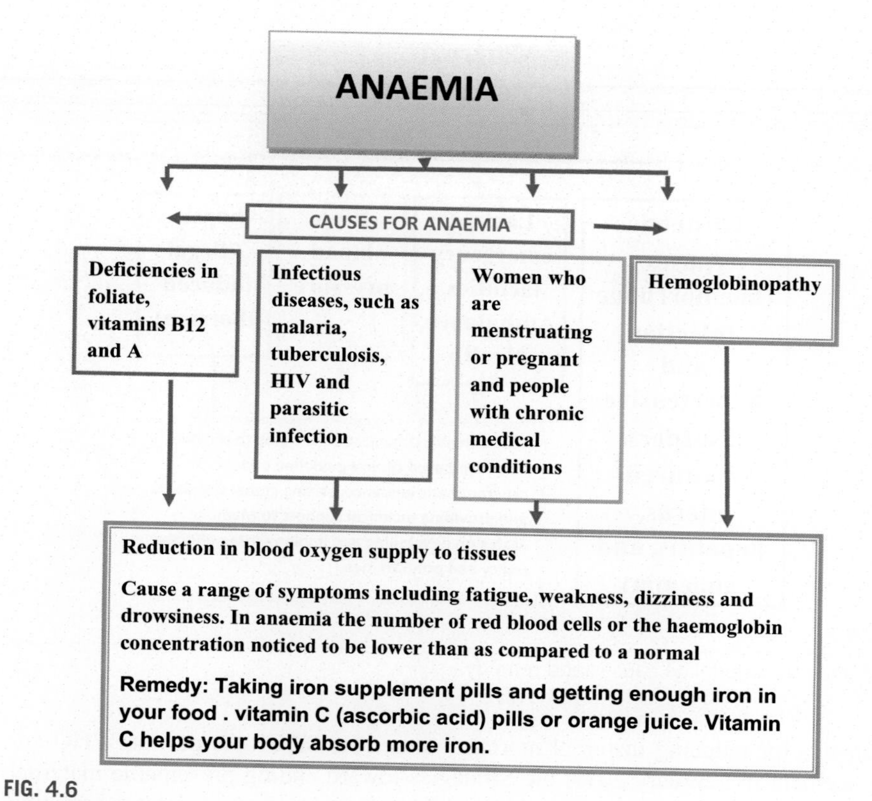

FIG. 4.6

The problem of anemia: causes and possible remedies.

the end of 2015. However, despite focused efforts, the reduction in maternal health targets in developing regions was not significantly achieved by 2015. After the completion of MDGs by the end of 2015, the United Nations officially announced an ambitious project with the title Sustainable Development Goals (SDGs), with 17 subtargets to eradicate poverty and hunger and achieve the universal health goal by the end of 2030. All the targets of MDGs were also taken into consideration while setting the SDGs program at global level. The 17 Sustainable Development Goals (SDGs) are a shared vision of humanity and a social contact between the world's leaders and people. The SDG3 subtarget of the SDGs seeks to ensure health and well-being for all, including a strong commitment to end the epidemics of AIDS, tuberculosis, malaria, and other communicable diseases by 2030. It also aims to achieve universal health coverage, and provide access to safe and effective medicines and vaccines for all. SDG3 ensures healthy lives on a priority basis, and promotes well-being for all at all ages in the SDGs agenda through 2030. In February 2015, the World Health Organization published: Strategies toward Ending Preventable Maternal Mortality (PMM). It mainly covers global targets and

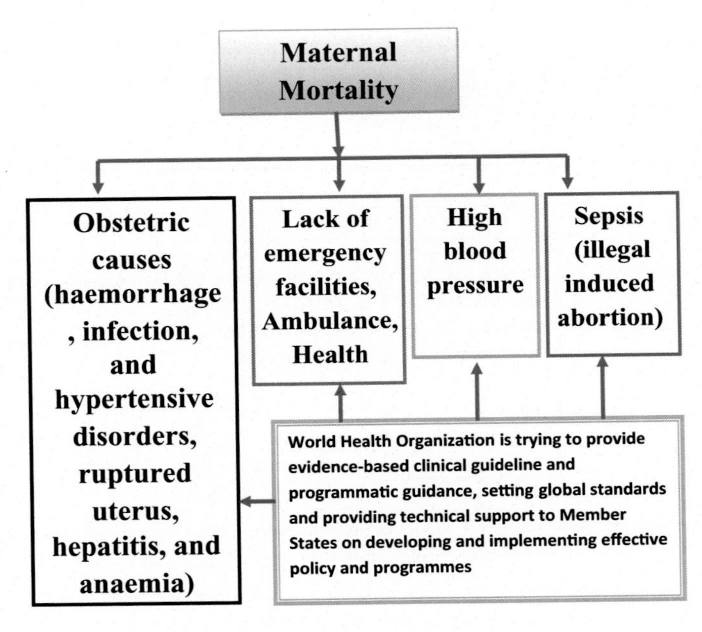

FIG. 4.7

MM (maternal mortality) causes and remedy.

strategies for reducing maternal mortality under the SDGs. As per the WHO announcement in February 2015, the Strategies toward ending preventable maternal mortality (EPMM) report outlined the global targets and strategies for reducing maternal mortality under the Sustainable Development Goals (SDGs).

In developed countries, severe complications of pregnancy and maternal deaths are rare. This is due to availability of emergency care and access to infrastructure facilities. With proper medical care, most of the problems that come about during pregnancy, childbirth, and the postpartum period can be treated or even prevented.

4.6.5.2 Facts and figures

Globally, maternal mortality is a great concern. About 295,000 women died during and following pregnancy and childbirth in 2017. The vast majority of these deaths (94%) occurred in low-resource settings, and most could have been prevented [76].

Almost all maternal deaths are preventable, yet about 86% (254,000) global estimated maternal deaths in 2017 occurred in Sub-Saharan Africa. During the same period, maternal death of Southern Asia was about one-fifth (58,000). The Southern Asia achieved about 60% declines in maternal mortality rate (MMR) over the value of the year 2000 (from an MMR of 384 down to 157). At the same time the MMR was reduced to half in four other sub-regions (Central Asia, Eastern Asia, Europe, and Northern Africa).

In some areas of the world, the high rate in MMR is mainly due to inequalities in access to quality health services and highlights the gap between rich and poor. The MMR in low-income countries in 2017 was 462 per 1,000,000 live births vs 11 per 100,000 live births in high-income countries.

In 2017, on the basis of high MMR and Fragile State Index, 15 countries (South Sudan, Somalia, Central Africa Republic, Yemen, Syria, Sudan, the Democratic Republic of Congo, Chad, Afghanistan, Iraq, Haiti, Guinea, Zimbabwe, Nigeria, and Ethiopia) were considered as "high alert," and these 15 countries had MMRs in 2017 ranging from 31 (Syria) to 1150 (South Sudan).

The risk of maternal mortality is highest for adolescent girl under 15 years old and complications in pregnancy and childbirth are higher among adolescent girl aged 10–19 (compared to women aged 20–24) [77,78].

4.6.5.3 WHO response

Improving maternal health is one of the key priorities. The WHO is trying to provide evidence-based clinical guideline and programmatic guidance, setting global standards and providing technical support to member states on developing and implementing effective policy and programs:

- Developing awareness and providing quality of reproductive, maternal, and newborn healthcare services.
- Ensuring successful implementation of universal health coverage, and linking it to reproductive, maternal, and newborn healthcare.
- Improving awareness of various aspects of maternal mortality, reproductive and maternal motilities, and related disabilities.
- Surveying and collecting evidence-based data to treat women and girls accordingly.
- Ensuring reliability on data interpretation and analysis to promote quality care on an equality basis.

4.6.6 Anthropometry

Four anthropometry measures are commonly used in expressing healthcare measures: weight, height, waist circumference (waist), and hip circumference. Additionally, two quotients derived from these measures, body mass index (BMI) weight kg/height^2 m^2 and waist-to-hip ratio (west/hip) are often used.

Anthropometry analysis is an essential key tool for understanding nutritional impact and assessment to determine malnutrition, overweight, obesity, muscular mass loss, fat mass gain, and adipose tissue redistribution. Anthropometric indicators are used for diagnosis of old age chronic diseases and to guide medical intervention.

4.6.6.1 Anthropometric status indicators

Commonly, stunting (H/A), wasting (W/H), underweight (W/A), and mid-upper arm circumference (MUCA) in children under 5 years of age, and body mass index (BMI) in adults are used in clinical analysis to understand the deficiency of nutrition in the body.

4.6.6.2 Height for age (H/A)

Stunting is a biological sign used to describe a condition in which children are unable to gain sufficient height in relation to their age. This is mainly due to the cumulative effects of under nutrition and infections, since and even before birth. Stunting is also the result of long-term nutritional deficiency (chronic malnutrition), and resulted in delayed mental development, frequent illness, poor school performance, and reduced innovative capacity. This measure can therefore be interpreted as indication of poor environmental conditions or long-term restriction of a child's growth potential. Stunting is very sensitive to socioeconomic inequalities, also.

4.6.6.3 Weight for height

During the growth and development period, a child is supposed to achieve weight for height (W/H). It is an indicator of current nutritional status. Wasting may be the consequence of frequent starvation due to some cultural belief or practice, or due to chronic disease.

4.6.6.4 Weight for age

The term "underweight" is use to describe a condition where a child weight is less than expected of her/his age.

Body mass index (BMI) is generally not applicable to adult populations, and is the same for both genders. BMI is not an accurate predictor of health because it does not account for body fat percentage or body fat distribution. The BMI may be a variable term with reference to geographic location of population. BMI may not correspond to the same degree of fitness in different populations, in part because of different body proportions. The health risks associated with increasing BMI are continuous, but the interpretation of the BMI grading in relation to risk may differ for different population.

The body mass indexes (BMIs) are age-dependent for adult populations of a specific locality, and are applied the same way to both males and females. Under the Global Nutrition Monitoring Framework, the proportion of underweight in women aged 15–49 years and of overweight in women aged 18 years or more are considered as intermediate outcome indicators. Adult overweight is also included in the NCD (Noncommunicable Diseases) Global Monitoring Framework and in the WHO Global Reference List of 100 Core Health Indicators.

4.6.6.5 How are indicators defined?

BMI is simply expressed in terms of weight to height, which is a measurement of underweight, overweight, and obesity in adults. It is expressed as the weight in kilograms divided by the square of the height in meters (kg/m^2). For example, an adult who weighs 58 kg and has a height of 1.7 m will have a BMI of 20.1, where BMI = 58 kg (1.70 m × 1.70 m) = 20.1. BMI values indicate the following:

BMI < 17.0: moderate and severe thinness.

BMI < 18.5: underweight.

BMI < 18.5–24.9: normal weight.

BMI ≥ 25.0: overweight.
BMI ≥ 30.0: obesity.

4.7 Initiatives in maternal healthcare

A declination in the rate of maternal mortality was observed as 44% between 1990 and 2015, thanks to the Millennium Development Goals program being initiated by the UN to eradicate poverty and hunger and implement universal coverage of health for a sustainable healthy life for all. After a service of 25 years, in September 2015 the noble era of the Millennium Development Goals (MDGs) came to end with memorable results for the well-being of humanity. Reduction in maternal mortality was also one of the key issues undertaken by the MDGs program. MDGs program was able to bring dawn the maternal mortality rate by 44% from 1990 to 2015. In spite of sincere efforts, the MDGs could not reach 75% reduction in maternal mortality rate, as targeted. This was mainly due to the socioeconomic status of many developing countries, which couldn't put effort into monitoring maternal health during the pregnancy period. So, maternal mortality remains unacceptably high, with approximately 303,000 maternal deaths occurring each year, with the largest burden in Sub-Saharan Africa and Asia [79].

The Sustainable Development Goals (SDGs), adopted in late 2015, to action an end to poverty, protect the natural world, and achieve habitable lives and universal health coverage all over the world. In this connection, a comprehensive paper, Strategies toward Ending Preventable Maternal Mortality (EPMM), was also launched in February 2015 to support achievement of the SDG global targets [80,81].

The SDGs consist of 17 goals, out of which, minimizing the maternal mortality ratio (MMR) is one of the priority issues. The SDG3, Target 3.4, is to minimize the maternal mortality rate to achieve a global average of 70 per 100,000 live births by 2030. Currently, progress is being made in many places, but, overall, action to meet the goals is not yet advancing as per the expected progress.

Factually, much earlier than the implementation of SDGs, at the beginning of January 2013, the Ending Preventable Maternal Mortality (EPMM) Working Group, led by the World Health Organization (WHO) with support from partner organizations, achieved multistakeholder consensus on goals for maternal health and survival from 2015 to 2030.

The overall programs of the EPMM for targeting maternal mortality reduction at the global and country level are:

- The global average maternal mortality ratio (MMR) should reach less than 70 maternal deaths per 100,000 live births, by the end of 2030.
- It is committed by all the members from all countries that, by 2030, every country should bring down the maternal mortality ratio by at least two thirds from 2010 baseline, and no country should have a ratio higher than 140 deaths per 100,000 live births.
- All countries are tasked with achieving equity in MMR among subpopulations.

The EPMM global MMR target was incorporated into the Sustainable Development Goals (SDGs) adopted by member states and launched in late 2015 [4]. Subsequently, SDGs form the basis of a new global development agenda, which is broad and comprehensive, covering a wide range of social, economic, and environmental goals. The EPMM global target was also included in the updated UN Global Strategy.

A wide range of civil society leaders and different voluntary organizations were invited to take active part to promote and accelerate progress on the Sustainable Development Goals, and extend their cooperation to strengthen efforts to reach the people of different geographical locations and involve them in adapting and understanding how SDGs can increase universal well-being.

4.7.1 Different global agencies' work for health

4.7.1.1 World Health Organization

The World Health Organization (WHO) is a specialized agency of the United Nations that works for various aspects of public health. The WHO is committed to "the attainment by all peoples of the highest possible level of health." Its headquarters are in Geneva, Switzerland. It has six regional offices and 150 field offices worldwide.

The WHO covers a wide range of issues related to universal healthcare, monitoring public health risks, coordinating responses to health emergencies, and promoting health and well-being. Apart from these services, the WHO also provides technical information and guidelines to assist developing countries, sets international health standards, and surveys data on global health issues. The WHO also serves as a forum for discussions of health issues.

The WHO is a world-leading organization, achieving landmark services like controlling polio, eradication of smallpox, and vaccination for many life-threatening diseases. Its current priorities include communicable diseases, particularly HIV/AID, COVID-19, malaria, Ebola, and tuberculosis; noncommunicable diseases such as heart disease and cancer; healthy diet, nutrition, and food security; occupational health; and substance abuse. In addition, the WHO has made effective strategies a high priority in dealing with maternal and child health (MCM) services and integrating vertical programs to give attention to care during labor and delivery, which is the most critical period in life for complications.

4.7.1.2 United Nations fund for Population Activities (UNFPA)

United Nations Fund for Population Activities (UNFPA) is a UN agency aimed at improving reproductive and maternal health worldwide. The agency was established in 1969 under the administration of the United Nations Development Fund. In 1987, its name was changed to United Nations Population Fund.

Its main responsibilities are to undertake issues on developing national healthcare strategies and protocols, develop access to birth control, and to bring awareness against child marriage, gender-based violence, obstetric fistula, and female genital mutilation. The work of UNFPA is across four geographic regions: the Arab

States and Europe, Asia and the Pacific, Latin America and the Caribbean, and Sub-Saharan Africa. It is a founding member of the United Nations Development Group, a collection of UN agencies and programs focused on fulfilling the Sustainable Development Goals.

4.7.1.3 United Nations Development Programme

The United Nations Development Programme's mandate is to eradicate poverty and develop democratic governance, the rule of law, and inclusive institutions. It provides countries with knowledge, experience, and resources to help people build a better life.

It is a global development network of the United Nations. The United Nation Development Programme (UNDP) operates internationally in 177 countries to promote the Sustainable Development Goals (SDGs) and to undertake works on poverty reduction and HIV/AIDS in association with local governing bodies. The UNDP is funded entirely by voluntary contributions from UN member states.

4.7.1.4 World Bank

The World Bank was established in 1944, as proposed in the Bretton Woods Conference, along with the International Monetary Fund (IMF). The president of the World Bank is traditionally an American. The World Bank and the IMF are both based in Washington DC, and work in close association with each other.

The World Bank funds temporary loans to low-income countries that could not obtain loans commercially. The Bank may also make loans and demand policy reforms from recipients. In this connection, for a country to be associated with the World Bank for any financial assistance, it has to develop strategies in cooperation with the local government and any interested stakeholders and may rely on analytical work performed by the Bank or other parties. The World Bank, in association with the International Finance Corporation (IFC) and Multilateral Investment Guarantee Agency (MIGA), provides a suite of financial products and technical advice and analysis to address development challenges, helping countries find solutions to achieve sustainable and inclusive development. The World Bank also finances government programs of developing countries to support achievement of country development objectives, and supports policy and institutional reforms of national and subnational governments by providing budget financing and global expertise. The following are some of the important fields for which World Bank finance for development and sustainability is available.

i) Extreme poverty and hunger eradication

Since the implementation of the Millennium Development Goals (MDGs) followed by the Sustainable Development Goals (SDGs), extreme poverty fell from almost a third to less than a fifth. The World Bank Group is committed to fighting against poverty in all its dimensions on the basis of the latest evidence and analysis to help governments to develop sound policies that can help the poorest in every country, and focus on investments in areas that are critical to improving lives.

ii) Universal primary education

About 258 million children and youths are out of school, according to UIS (the UNESCO Institute for Statistics) data for the school year ending in 2018. The total includes 50 million children of primary school age, 62 million of lower secondary school age, and 138 million of upper secondary age.

The World Bank Group is the largest financier of education in the developing world. It works on education programs in more than 80 countries and is committed to helping countries reach SDG4, which calls for access to quality education and lifelong learning opportunities for all by 2030.

iii) Promoting gender equality

The World Bank Group asserts that no country, community, or economy can successfully achieve its potential or challenge the 21st century by ignoring the full and equal participation of women and men, girls and boys. So, the World Bank Group (WBG) promotes gender equality in developing countries through lending, grants, knowledge, and analysis and policy dialogue. The Bank is committed to ensure good levels of gender mainstreaming in its operations and that all Country Assistance Strategies are gender-informed.

iv) Universal health coverage

Achieving universal health coverage (UHC) is a core target of the Sustainable Development Goals (SDGs). The World Bank Group (WBG) is supporting countries' efforts toward this goal and to provide quality affordable health services to everyone regardless of their ability to pay by strengthening primary healthcare systems and reducing the financial risks associated with ill health and increasing equality.

In continuation of this, with global confrontation with the COVID-19 pandemic in September 2020, the World Bank declared a $12 billion plan to supply "low- and middle-income countries" with a vaccine once it is approved. About 2 billion people will be benefited with this mega donation.

v) Developing environmental sustainability

The World Bank Group provides environmental expertise, technical assistance, and financing to help low- and middle-income countries manage land, sea, and freshwater natural resources in a sustainable way to help create jobs, improve livelihoods, and enhance ecosystems.

vi) Development of a global partnership for development

On the basis of a country's (particularly of developing country) demand, numerous projects donated by various international agencies and finance for up gradation or sustainability of the projects need proper guidelines and procedures designed to protect the project and ensure that it aids the poor.

The World Bank Group works, in association with other international institutions and donors, civil society, and professional and academic associations, to improve the coordination of aid policies and practices in countries, at the regional level and at the global level.

4.7.1.5 Swedish International Development Cooperation Agency

The Swedish International Development Cooperation Agency (SIDA) is a government agency of the Swedish Ministry for Foreign Affairs. SIDA works on poverty eradication and improving basic living standards. Mainly, SIDA focuses on five areas: democracy, equality, and human rights; economic development; health and social development; sustainable development; and peace and security. As asserted by SIDA, all these areas are important in reducing world poverty and creating fair and sustainable development.

SIDA believes in work in partnership with private sector organizations for developing economic, social, and environmentally sustainable development throughout their activities, production processes, and value chains. SIDA's intension is to strengthen the partnership programs with the private sector to promote trade, technology transfer, and problem solving.

SIDA provides help to ongoing multicountry programs on HIV/AIDS prevention and impact mitigation in the world of work in Sub-Saharan Africa.

4.7.1.6 United Nations food and Agriculture Organization

On October 16, 1945, 42 countries gathered in Quebec, Canada, to create the Food and Agriculture Organization (FAO) of the United Nations. The main focus of the FAO is to eradicate hunger and improve nutrition and food security. Mainly, FAO works from Rome and Italy by maintaining regional and field offices around the world, covering 130 countries. In coordination with the government of different countries and other volunteer nongovernment organizations, it coordinates to improve and develop agriculture, forestry, fisheries, and land and water resources. It is also involved in conducting various training programs on short and long term bases to improve and update technical knowledge on production quality improvement, and collect data on agricultural output, production, and development.

4.8 Health insurance policy at country level

Health insurance or medical insurance is a type of provision for covering the whole or part of the risk of a person incurring medical expenses. The modality of health insurance mainly depends on a state government. However, as the basic functioning rule of operating health insurance is by estimating the overall risk of health risk and health system expenses over the risk pool, an insurance can develop a routine finance structure, such as a monthly premium or payroll tax, to provide money to pay for the healthcare benefits specified in the insurance agreement. The benefit is administered by a central organization, such as a government agency, private business, or not-for-profit entity.

4.8.1 Australia

Australia operates universal public health insurance, known as Medicare. In 1974, the new parliament of Australia passed the health care legislation by establishing free hospital care and subsidized private care. However, following a change in government in 1975, access to free health care services was limited to retired persons who met stringent means tests. Again, since another change of government in 1984, the current Medicare system provides free public hospital care and substantial coverage for physicians' services and pharmaceuticals for Australian citizens, residents with permanent visas, and New Zeeland citizens following their enrolment in the program and conformation of identity. Restricted access is provided to citizens of certain other countries through formal agreements.

4.8.2 Brazil

In 1988, Brazil created health insurance on the basis of "health as universal right." The Brazilian health system is known as SUS (*Sistema Unico de Saude*). Brazil provides free universal access to medical care to anyone legally living in the country.

Several hundred insurance firms in Brazil offer four principal types of medical plans: private health insurance, prepaid group practice, medical cooperatives, and company health plans. The costs of private insurance in Brazil vary accordingly to the provider, coverage, and region. They also have varying terms and conditions.

4.8.3 Canada

Canada has a decentralized, universal, publicly funded health care system called Canadian Medicare. Canada's insurance system is funded by the country's 13 provinces and territories. Each has its own insurance plan, and each receives cash assistance from the federal government on a per-capita basis. Canadian Medicare provides coverage for approximately 70% of Canadians' healthcare needs, and the remaining 30% is paid for through the private sector. The 30% typically relates to services not covered or only partially covered by Medicare, such as prescription drugs, eye care, and dentistry.

4.8.4 China

China largely achieved universal insurance coverage in 2011 through three public insurance programs: (i) Urban Employee Basic Medical Insurance, mandatory for urban residents with formal jobs, was launched in 1998; (ii) the voluntary Newly Cooperative Medical Scheme was offered to rural residents in 2003; and (iii) the voluntary Urban Resident Basic Medical Insurance was launched in 2007 to cover urban residents without formal jobs, including children, the elderly, and the self-employed.

In 2016, the Newly Cooperative Medical Scheme and Urban Resident Basic Medical Insurance scheme were merged by a joint action committee of China's central government, and the State Council. This step was taken to reduce administrative costs.

In order to minimize financial burden, in 2016, the Newly Cooperative Medical Scheme and Urban Resident Basic Medical Insurance scheme were merged by a joint action committee of China's central government, and the State Council. In 2011, approximately 95% of the Chinese population was covered under one of the three medical insurance schemes.

4.8.5 Denmark

The citizens of Denmark are automatically enrolled in a publicly financed healthcare system, which is largely free at the point of use. Registered immigrants and asylum-seekers are also covered, while undocumented immigrants have access to acute-care services through a voluntary copayments system. Under both the insurance options, access to hospital requires a referral.

4.8.6 United Kingdom

Since 1948, with the creation of the National Health Services (NHS), the UK has been under the coverage of a universal health system. The NHS was established under the National Health Service Act of 1946. The main salient feature of the NHS is free health service, replacing voluntary insurance and out-of-pocket payment. The free health care is aimed at wider welfare reform designed to eliminate unemployment, poverty, and illness, and to improve education.

As prescribed in the NHS Constitution, all eligible persons have the right to access health care without discrimination and within certain time limits for certain categories, such as emergency and planned hospital care. All UK citizens and legal residents are automatically entitled to NHS care, still largely free at the point of use, as are nonresidents with a European Health Insurance Card. For non-European visitors or undocumented immigrants, only treatment in an emergency department and for certain infectious diseases is free.

4.8.7 France

Since 1945, France has been enjoying universal health coverage by extending statutory health insurance (SHI) to all employees.

In 2000, the *Couverture Maladie Universelle* (Universal Health Coverage) or CMU was created for residents not eligible for SHI. After the implementation of CMU, less than 1% of residents were left without baseline coverage.

In January 2016, SHI eligibility was universally granted under a universal health protection law to fill in the few remaining coverage gaps. The law also replaced and simplified the existing system by providing systematic coverage to all French residents.

4.8.8 India

The right to health is a fundamental right to every citizen of India under Article 21 of the Constitution of India. In India, each person is supposed to receive free universal access to healthcare services. However, healthcare in India has been chronically inadequate. Historically, there have been several government-funded health insurance schemes intended to cover selected beneficiaries at state level or national level. One of the most important healthcare policies for lower-income people in India was implemented in the name of National Health Insurance Program

(*Rashtriya Swashthya Bima Yojana* or RSBY) in 2008. As of 2016, some 41 million families were enrolled in RSBY. But, due to administrative operational hurdles, the scheme output has not reached the expected result. In 2018, the Government of India launched *Ayushman Bharat PM-JAY*. *Ayushman Bharat Yojana*, also known as the *Pradhan Mantri Jan Arogya Yojana* (PMJAY) is a scheme that aims to help economically vulnerable Indians who are in need of healthcare facilities.

PMJAY subsumed the two other centrally sponsored schemes: the *Rashtriya Swaathya Bima Yojana* (RSBY), which was launched in 2008 and insured families of informal sector workers below the poverty line (BPL) for up to Rs 30,000 ($405.15), and the Senior Citizen Health Insurance Scheme (SCHIS), which was launched in 2016,and provides an additional Rs 30,00 ($405.15) for every senior citizen in families eligible for RSBY. However, 3 years since the *Ayushman Bharat-Pradhan Matri Jan Arogya* began, it is yet to meet one of its main objectives of seamless, paperless, and cashless access to services at the point of care. Evidence indicates that the scheme has not significantly reduced out-of-pocket payment.

Due to inefficacy in operating public insurance schemes and low acceptability of commercial insurance schemes, only about 37% of the people were covered by any form of health coverage in 2017–18. The main barriers for such inefficacy of low public healthcare insurance acceptability by the public include long wait times in hospital, the perceived low quality of public health services, and substantial workforce and infrastructure shortfalls.

Although, India is the second most populous country, on the basis of gross domestic product (GDP) the health expenditure of India is among the lowest at 3.54% of GDP. Of this, the share of public expenditure is about 1.28% of GDP, indicating that healthcare in India is largely private owned. So, the overall situation of health financing is an important social security measure, and enabling access generates demand for improved healthcare and ensuring financial risk protection. So, in order to develop an effective health insurance market it is necessary to reduce the "out-of-pocket" money of the public which is about 62% of total expenses towards health care maintenances.

4.8.9 United States of America

The United States does not follow a universal health insurance policy system. Health insurance is covered by the private sector, with different commercial brand names. In 2018, about 92% of the population used the facilities of health insurance being floated by different commercial agencies. About 27.5 million people, or 8.5% of the population are uninsured.

In 1920, employer-sponsored health insurance was introduced, which gained public acceptability immediately after World War II when the government imposed wage controls and declared fringe benefits such as health insurance as tax-exempt. In 2018, about 55% of the population were covered by an insurance policy under employer-sponsored insurance. In 1965, the first public insurance programs with the names "Medicare" and "Medicaid" were introduced. Medicare is applicable for persons aged 65 and older. In later stages, persons under 65 but suffering from long-term disabilities or end-stage renal disease became eligible. In Medicaid, health insurance programs first gave states the option to receive federal matching funding for providing healthcare services to low-income families, the blind, and individuals with disabilities. Later, the coverage was gradually made mandatory for low-income pregnant women and infants, and children up to 18. Currently, 17.9% of Americans use Medicaid. As of 2019, more than two-thirds of Medicaid beneficiaries were enrolled in managed care organizations.

In 1997, the Children's Health Insurance Program, or (CHIP), was created as a public, state-administered program for children in low-income families that earn too much to qualify for Medicaid but that are unlikely to be able to afford private insurance. Presently, the program covers 9.6 million children. In some states it is applied as an extension of Medicaid; in other states, it is a separate program.

4.8.10 Switzerland

Currently, Switzerland operates a universal health coverage system, as stated in the Health Insurance Law in 1994. It is based on a private insurance model with the objectives to strengthen equality by introducing universal coverage and subsidies for low-income households; and to provide a high standard of access for healthcare, even increasing the cost of healthcare related aids.

4.8.11 Sweden

Sweden provides universal health coverage to all registered citizens, automatically. All patients belonging to the European Union are entitled to universal health coverage, subject to condition that the countries should be from European Economic Area, and the nine other countries with which Sweden has agreements. Asylum-seeking and undocumented children have the right to healthcare services. The universal health care systems of Sweden also equally take care of other foreign immigrants to provide health care services for quality life under universal health care insurance scheme. The cost measurement of universal health coverage should be to improve health and quality of life.

4.8.12 Italy

Italy provides universal health coverage under Italy's National Health Services (*sevizio sanitaria nazionale* or SSN) established through legislation in 1978. The universal health coverage is automatically applied to all citizens and legal foreign residents. Temporary visitors are supposed to meet healthcare costs from their own pocket.

4.8.13 Israel

In 1995, Israel's National Health Insurance (NHI) was approved, providing universal coverage for all citizens and permanent residents. The Israeli law states that: "Health insurance ... shall be based on the principle of justice, equality and mutual assistance."

All citizens have right to receive all services included in the benefit package that is mandated by the government, subject to medical discretion. Residents have right to receive medical services at a reasonable quality level, within a reasonable period of time, and reasonable distance from their home. There is no formal definition of "reasonable," and there is no penalty for health plans that fail to comply.

Those serving in the army (as they receive health care directly from the army); inmates, who receive care from the Israel Prison Service; documented and undocumented foreign workers, whom employers are required to enroll in private insurance programs; and undocumented migrants, temporary residents, and tourists also get opportunity to avail universal health care scheme for maintenance of quality health.

4.8.14 Germany

The world's first social health insurance system "Chancellor Otto von Bismarck's Health Insurance Act" was established in 1883 in Germany. At the beginning, the health insurance system was limited to blue-collar workers. In 1885, 10% of people were insured and entitled to cash benefits in case of illness, death, or childbirth. In 2007, the current universal health insurance was adopted for all citizens and permanent residents. Currently, universal health coverage is for the entire population, along with a generous benefit package.

Presently, two systems of health insurance exist: statutory health insurance (SHI), consisting of competing, not-for-profit, nongovernmental health insurance plans known as sickness funds; and private health insurance. For long-term care, German has statutory long-term care insurance (LTCL). Hospitals and physicians treat all patients regardless of whether they have SHI or private insurance.

4.8.15 Japan

Japan has statutory health insurance system (SHIS), which covers 98.3% of the population. The remaining 1.7% of people are covered by a separate public social assistance program. Noncitizens have to enroll in an SHI plan to receive health coverage; but undocumented immigrants and visitors do not have access to health coverage. Two types of mandatory provision exist in the SHI system: (i) employment-based insurance plans, which cover about 59% of the population and (ii) residence-based insurance plans, which include Citizen Health Insurance plans for nonemployee individuals aged 74 and under and Health Insurance for the Elderly Plans, which automatically cover all adults aged 75 and older.

4.8.16 Netherlands

In 1941, the Netherlands adopted a model based on the first health insurance system: the German Bismarck model of public and private health insurance. However, the success in health insurance was not satisfactory, and as a result, in 2006, the Health Insurance Act merged the traditional public and private insurance markets into one universal social health insurance program underpinned by private insurance and mandatory coverage. All residents (and nonresidents who pay Dutch income tax) must purchase statutory health insurance from private insurers. Children under 18 are automatically covered, while adults choose a policy on an individual basis. Active members of the armed forces (who are covered by the Ministry of Defense) are exempt.

Undocumented immigrants cannot purchase health insurance and have to pay for most treatments out-of-pocket (excluding acute care, obstetric services, and long-term care). However, some alternate provisions are available to cover undocumented immigrants who are unable to pay. Political asylum-seekers fall under a separate, limited insurance plan. Permanent residents of the Netherlands who have been in the country for more than 3 months are obliged to purchase private insurance. Short-term visitors are required to purchase insurance so long as they intend to stay, as per rule.

4.8.17 New Zealand

At the beginning of 1938, as per the Social Security Act., a consensus was developed among the people of New Zealand for the provision of the population's healthcare. Without any further delay, the government implemented universal health coverage under which every citizen of New Zealand is entitled to public healthcare without any discrimination. However, in practice, the health insurance coverage varies by the income, need, location, and type of service.

4.8.18 Norway

Norway provides universal health and social insurance coverage, under the title the National Insurance Scheme (NIS), which is currently regulated by the 1997 National Insurance Act and the 1999 Patient Right Act.

The establishment of universal coverage has a long history in Norway. Political and social movements began advocating for universal social and healthcare insurance around 1900. The Act of Health Insurance, covering employees as well as their families, came into force in 1909. Membership was mandatory for low-income employees; others could opt in. The coverage was twofold: healthcare and guaranteed basic income in cases of income loss due to ill health. In 1956, the system was converted into a universal and mandatory right for all citizens.

References

[1] Verbeke W, Ward RW. Consumer interest in information cues denoting quality, traceability and origin: an application of ordered probit models to beef labels. Food Qual Prefer 2006;17:453–67.

[2] Kolodinsky J. Persistence of health labeling information asymmetry in the United States: historical perspectives and twenty-first century realities. J Macromark 2012;32:193–207.

[3] Cerri J, Testa F, Rizzi F. The more I care, the less I will listen to you: how information, environmental concern and ethical production influence consumers' attitudes and the purchasing of sustainable products. J Clean Prod 2018;175:343–53.

[4] Mayounga AT. Antecedents of recalls prevention: analysis and synthesis of research on product recalls. Supply Chain Forum Int J 2018;19(3).

[5] Kolodinsky J. Persistence of health labelling information asymmetry in the United States: historical perspectives and twenty-first century realities. J Macromark 2012;32:193–207.

[6] Constitution of the World Health Organization. World Health Organization: basic documents. 45th ed. Geneva: World Health Organization; 2005.

[7] Hooper L, Abdelhamid A, Bunn D, Brown T, Summerbell CD, Skeaff CM. Effects of total fat intake on body weight. Cochrane Database Syst Rev 2015;(8), CD011834.

[8] Anon. Diet, nutrition and the prevention of chronic diseases: report of a Joint WHO/FAO Expert Consultation. WHO Technical Report Series, No. 916. Geneva: World Health Organization; 2003.

[9] Anon. Fats and fatty acids in human nutrition: report of an expert consultation. FAO Food and Nutrition Paper 91. Rome: Food and Agriculture Organization of the United Nations; 2010.

[10] Nishida C, Uauy R. WHO scientific update on health consequences of trans fatty acids: introduction. Eur J Clin Nutr 2009;63(Suppl 2):S1–4.

[11] Anon. Guidelines: saturated fatty acid and *trans*-fatty acid intake for adults and children. Geneva: World Health Organization; 2018 [Draft issued for public consultation in May 2018].

[12] Anon. REPLACE: an action package to eliminate industrially-produced *trans*-fatty acids. WHO/NMH/NHD/18.4. Geneva: World Health Organization; 2018.

[13] Anon. Guideline: sugars intake for adults and children. Geneva: World Health Organization; 2015.

[14] Anon. Guideline: sodium intake for adults and children. Geneva: World Health Organization; 2012.

[15] Anon. Comprehensive implementation plan on maternal, infant and young child nutrition. Geneva: World Health Organization; 2014.

[16] Anon. Global action plan for the prevention and control of NCDs 2013–2020. Geneva: World Health Organization; 2013.

[17] World Health Organization. Infant and young child feeding; 2021. https://www.who.int>newsroom>Fact sheet>Detail.

[18] Azevedo WF, Diniz MB, Fonseca ES, Azevedo LM, Evangelista CB. Complications in adolescent pregnancy: systematic review of the literature. Einstein (Sao Paulo) 2015;13(4):618–26. https://doi.org/10.1590/S1679-45082015RW3127.

[19] Bacchus LJ, Ranganathan M, Watts C, Devries K. Recent intimate partner violence against women and health: a systematic review and meta-analysis of cohort studies. BMJ Open 2018;8. https://doi.org/10.1136/bmjopen-2017-019995, e019995.

[20] Bilano VL, Ota E, Ganchimeg T, Mori R, Souza JP. Risk factors of pre-eclampsia/eclampsia and its adverse outcomes in low- and middle-income countries: a WHO secondary analysis. PLoS One 2014;9(3). https://doi.org/10.1371/journal.pone.0091198, e91198.

[21] Buzi RS, Smith PB, Kozinetz CA, Peskin MF, Wiemann CM. A socioecological framework to assessing depression among pregnant teens. Matern Child Health J 2015;19(10):2187–94. https://doi.org/10.1007/s10995-015-1733-y.

[22] Ellsberg M, Jansen HA, Heise L, Watts CH, Garcia-Moreno C. Intimate partner violence and women's physical and mental health in the WHO multi-country study on women's health and domestic violence: an observational study. Lancet 2008;371(9619):1165–72. https://doi.org/10.1016/S0140-6736(08)60522-X.

[23] Darroch J, Woog V, Bankole A, Ashford LS. Adding it up: costs and benefits of meeting the contraceptive needs of adolescents. New York: Guttmacher Institute; 2016.

[24] UNICEF. Ending child marriage: progress and prospects. New York: UNICEF; 2013.

[25] WHO. Global and regional estimates on violence against women: prevalence and health effects of intimate partner violence and non-partner sexual violence. Geneva: WHO; 2013.

[26] WHO, UNICEF, UNFPA, World Bank Group and the United Nations Population Division. Trends in maternal mortality: 1990 to 2015: Estimates by WHO, UNICEF, UNFPA, World Bank Group and the United Nations Population Division. Geneva: WHO; 2015. Filippi V, Chou D, Ronsmans C, et al. Levels and causes of maternal mortality and morbidity. In: Black RE, Laxminarayan R, Temmerman M, et al., editors. Reproductive, maternal, newborn, and child health: disease control priorities. 3rd ed. (vol. 2). Washington, DC: The International Bank for Reconstruction and Development/ The World Bank; 2016 Apr 5 [Chapter 3].

[27] Kozuki N, Lee A, Silveira M, et al. The associations of birth intervals with small-for-gestational-age, preterm, and neonatal and infant mortality: a meta-analysis. BMC Public Health 2013;13(Suppl. 3):S3.

[28] World Bank. Economic impacts of child marriage: global synthesis report. Washington, DC: World Bank; 2017.

[29] WHO. Preventing early pregnancy and poor reproductive outcomes among adolescents in developing countries. Geneva: WHO; 2011.

[30] UNESCO. International technical guidance on sexuality education: an evidence-informed approach for schools, teachers and health educators. Paris: UNESCO; 2009.

[31] UNESCO. Early and unintended pregnancy & the education sector: evidence review and recommendations. Paris: UNESCO; 2017.

[32] United Nations General Assembly. Resolution adopted by the general assembly on 25 September 2015: transforming our world: the 2030 agenda for sustainable development. New York: United Nations; 2015.

[33] Tayebi TT, et al. Relationship between revised graduated index (R-GINDEX) of prenatal care utilization & preterm labor and low birth weight. Global J Health Sci 2014;6:no. 3.

[34] Majrooh MA, et al. Coverage and quality of antenatal care provided at primary health care facilities in the "Punjab" province of "Pakistan". PLoS One 2014;9(11):e113390:1–8.

[35] Lincetto O, Mothebesoane-Anoh S, Gomez S, Munjanja S. Opportunities for Africa's newborns: practical data, policy, and programmatic support for Newborn Care in Africa. Geneva, Switzerland: World Health Organization; 2006. http://www.who.int/pmnch/media/publications/aonsectionIII_2.pdf.

[36] WHO. Antenatal care in developing countries: promises, achievements and missed opportunities; an analysis of trends, levels and differentials 1990–2001. Geneva, Switzerland: WHO; 2003.

[37] WHO. WHO recommendations on antenatal care for a positive pregnancy experience. Geneva, Switzerland: WHO; 2016. http://www.who.int/reproductivehealth/publications/maternal_perinatal_health/anc-positive-pregnancy-experience/en/.

[38] Tunçalp O, Pena-Rosas J, Lawrie T, et al. WHO recommendations on antenatal care for a positive pregnancy experience-going beyond survival. BJOG 2017;124(6):860–2.
[39] Chemir F, Alemseged F, Workneh D. Satisfaction with focused antenatal care service and associated factors among pregnant women attending focused antenatal care at health centers in Jimma town. BMC Res Notes 2014;7(1):1–7.
[40] Asefa F, Fekadu G, Taye A. Quality of antenatal care at Jimma medical center south West Ethiopia. Ethiop J Reprod Health 2020;12:1.
[41] Islam MM, Masud MS. Determinants of frequency and contents of antenatal care visits in Bangladesh: assessing the extent of compliance with the WHO recommendations. PLoS One 2018;13(9), e0204752.
[42] Akachi Y, Kruk ME. Quality of care: measuring a neglected driver of improved health. Bull World Health Org Policy Practice 2017;96(6):465–72.
[43] Health Information and Quality Authority. Guidance on developing key performance indicators and minimum data sets to monitor healthcare quality: February (Version 1.1). Dublin, Ireland: Health Information and Quality Authority; 2013.
[44] Mosadeghrad AM. Factors influencing healthcare service quality. Int J Health Policy Manag 2014;3(2):77–89.
[45] Nylenna M, et al. What is good quality of health care? Prof Prof 2015;5(1):1–15.
[46] Mainz J. Defining and classifying clinical indicators for quality improvement. International J Qual Health Care 2003;15(6):523–30.
[47] Yeoh PL, Hornetz K, Dahlui M. Antenatal care utilisation and content between low-risk and high-risk pregnant women. PLoS One 2016;11(3), e0152167.
[48] Beeckman K, Louckx F, Masuy-Stroobant G, Downe S, Putman K. The development and application of a new tool to assess the adequacy of the content and timing of antenatal care. BioMed Centr Health Serv Res 2011;11(213):1–10.
[49] Hijazi HH, et al. Determinants of antenatal care attendance among women residing in highly disadvantaged communities in northern Jordan: a cross-sectional study. Reprod Health 2018;15.
[50] Woyessa HA, Ahmed TH. Assessment of focused antenatal care utilization and associated factors in Western Oromia, Nekemte, Ethiopia. BMC Res Notes 2019;12:277.
[51] Abate TM, Salgedo BS, Bayou NB. Evaluation of the quality of antenatal care (ANC) service at higher 2 health center in Jimma, South west Ethiopia. Open Access Libr J 2015;2:e1398:1–9.
[52] Servan-Mori E, et al. Timeliness, frequency and content of antenatal care: which is most important to reducing indigenous disparities in birth weight in Mexico? Health Policy Plan 2016;31(4):444–53.
[53] Dansereau E, et al. Coverage and timing of antenatal care among poor women in 6 Mesoamerican countries. BMC Pregnancy Childbirth 2016;16:no. 234.
[54] Amoakoh-Coleman M, et al. Client factors affect provider adherence to clinical guidelines during first antenatal care. PLoS One 2016;11(6), e0157542.
[55] Muchie KF. Quality of antenatal care services and completion of four or more antenatal care visits in Ethiopia: a finding based on a demographic and health survey. BMC Pregnancy Childbirth 2017;17(300).
[56] Kyei-Nimakoh M, Carolan-Olah M, McCann TV. Access barriers to obstetric care at health facilities in sub-Saharan Africa—a systematic review. Syst Rev 2017;6:no. 110.
[57] Singh A, Kumar A, Pranjali P. Utilization of maternal healthcare among adolescent mothers in urban India: evidence from DLHS-3. PeerJ 2014;2, e592.
[58] Kalule-Sabiti L, Amoateng AY, Ngake M. The effect of socio-demographic factors on the utilization of maternal health care services in Uganda. Afr Pop Stud 2014;28(1):515–25.

[59] Joshi C, Torvaldsen S, Hodgson R, Andrew HA. Factors associated with the use and quality of antenatal care in Nepal: a population-based study using the demographic and health survey data. BMC Pregnancy Childbirth 2014;14(94).
[60] Ngongo N. Health system predictors of antenatal care compliance among rural Congolese women. Walden Dissertations and Doctoral Studies, Minneapolis, MN, USA: Walden University; 2016. https://scholarworks.waldenu.edu/dissertations/2038.
[61] Basha GW. Factors affecting the utilization of a minimum of four antenatal care services in Ethiopia. Obstetr Gynecol Int 2019;2019:1–6.
[62] Tura G. Antenatal care service utilization and associated factors in Metekel zone, Northwest Ethiopia. Ethiop J Health Sci 2011;19:no. 2.
[63] Getachew T, Abajobir AA, Aychiluhim M. Focused antenatal care service utilization and associated factors in Dejen and Aneded districts, Northwest Ethiopia. Prim Health Care 2014;4:170.
[64] Fagbamigbe AF, Idemudia ES. Barriers to antenatal care use in Nigeria: evidences from non-users and implications for maternal health programming. BMC Pregnancy Childbirth 2015;15:95.
[65] Ononokpono DN, Odimegwu C, Imasiku ENS, Adedini SA. Contextual determinants of maternal health care service utilization in Nigeria. Women Health 2013;53(7):647–68. https://doi.org/10.1080/03630242.2013.826319.
[66] NPC, ICF International. Nigeria demographic and health survey 2013. Abuja and Rockville: National Population Commission, Nigeria and ICF International; 2014.
[67] Odetola TD. Health care utilization among rural women of child-bearing age: a Nigerian experience. Pan Afr Med J 2015;20:151. https://doi.org/10.11604/pamj/2015.20.151.5845.
[68] Federal Ministry of Health. National Strategic Health Development Plan (NSHDP); 2010. p. 2010–5.
[69] Federal Government of Nigeria. Integrating primary health care governance in Nigeria (PHC under one roof). Abuja: Implementation Manual: National Health Care Development Agency; 2013.
[70] Simonds SK. Health education as social policy. Health Educ Monogr 1974;21:1–10.
[71] Pleasant A, McKinney J. Coming to consensus on health literacy measurement: an online discussion and consensus-gauging process. Nurs Outlook 2011;59(2):95–106.e1. https://doi.org/10.1016/j.outlook.2010.12.006. PMID 21402205.
[72] Atkinson RC, Jackson GB. Res Educ Reform 1992. https://doi.org/10.17226/1973. ISBN 978-0-309-04729-6.
[73] U.S. Department of Health and Human Services, Office of Disease Prevention and Health Promotion. National action plan to improve health literacy. Report, Washington, DC; 2010.
[74] Ratzan SC, Parker RM. Introduction. In: Selden CR, Zorn M, Ratzan SC, Parker RM, editors. National Library of Medicine current bibliographies in medicine: health literacy. Bethesda, MD: National Institutes of Health, U.S. Department of Health and Human Services; 2000. NLM Pub. No: CBM 2000-1.
[75] Vernon J, Trujillo A, Rosenbaum S, DeBuono B. Low health literacy: implications for national health policy. University of Connecticut; 2007.
[76] Trends in Maternal Mortality. 2000 to 2017: estimates by WHO, UNICEF, UNFPA, World Bank Group and the United Nations Population Division. Geneva: World Health Organization; 2019.
[77] Ganchimeg T, Ota E, Morisaki N, et al. Pregnancy and childbirth outcomes among adolescent mothers: a World Health Organization multicountry study. BJOG 2014;121(Suppl 1):40–8.

[78] Althabe F, Moore JL, Gibbons L, et al. Adverse maternal and perinatal outcomes in adolescent pregnancies: the global network's maternal newborn health registry study. Reprod Health 2015;12(Suppl 2):S8.
[79] WHO, UNICEF, UNFPA, World Bank Group and United Nations Population Division. Trends in maternal mortality: 1990 to 2015. Geneva: World Health Organization; 2015.
[80] WHO. Strategies toward ending preventable maternal mortality; 2015. Available at: http://who.int/reproductivehealth/topics/maternal_perinatal/epmm/en/. [Accessed 19 Aug 2016].
[81] Chou D, et al. Ending preventable maternal and newborn mortality and stillbirths. BMJ 2015;351, h4255. Available at: http://www.bmj.com/content/bmj/351/bmj.h4255.full.pdf.

CHAPTER

Social, economic, and health disparities of rural women 5

5.1 Women's health condition in developing countries

It is obvious that both women and men suffer thru poverty, gender discrimination, and socioeconomic problems, but still, women are provided with fewer resources to cope. In this world, women are the last to eat and are routinely trapped by their family circumstances, placing the well-being of every other member of the household before their own.

Most developing countries face lack of access to health risk management, and confront life-altering and life-threatening issues. In the case of pandemic infectious disease like COVID-19, women in rural communities and highly remote localities face an impossible situation with the family. Maternal mortality, female genital cutting, and child marriage hold back developing nations from making further progress.

It is high time to realize to give top most priority to women's health as they are the main pillar of strengthening family and social structure and function with particular reference to rural women. In order to resolve this great crisis we should intensify our attention on women's rights, to which they are entitled as human rights. Women's rights include the right to live free from violence and discrimination; to enjoy the highest attainable standard of physical and mental health; to be educated; to own property; to vote; and to earn equal pay [1,2].

So, we should pay maximum attention to stop maternal and neonatal mortality and global level. Therefore, various international bodies have undertaken the globally oriented program to achieve the Sustainable Development Goals within the projected time or before. In order to bring awareness to various issues in women's health, the world celebrates "International Women's Day."

5.2 Why celebrate International Women's Day?

March 08 is celebrated as "International Women's Day (IWD)" all over the world to raise awareness on issues related to women's health and well-being. Currently, it is a global holiday. In this connection global holiday means to allow individual to celebrate or commemorate an event related to wellbeing of women as a whole to bring social inclusion and harmony at global level. Its purpose is also to focus on the women's rights movement by bringing awareness of issues such as gender equality, reproductive rights, and violence and abuse against women (Fig. 5.1).

Healthcare Strategies and Planning for Social Inclusion and Development. https://doi.org/10.1016/B978-0-323-90447-6.00005-9

FIG. 5.1

Main motto of International Women's Day.

5.2.1 Historical movement on women's rights

5.2.1.1 International Women's Day in early 1900 and onwards

International Women's Day (IWD) has been observed since the early 1900s, when the rapid growth of industries was dramatically synchronized with population growth, and resulted in heavy migration of people, including women from rural communities to urban areas located near industries. Women at this time were increasingly feeling helpless, without the power or the right to vote. Women's oppression and inequality was spurring women to become more vocal and active in campaigning for change. In 1908, about 15,000 women raised their voice on the streets of New York, demanding shorter hours, better pay, and voting rights (Fig. 5.2).

In February 1909, the first National Women's Day (NWD) was celebrated by the Socialist Party of America. In response to this increasingly organized

FIG. 5.2

International Women's Day: protest regarding women's wages and women's voting rights on the streets of New York, 1908.

movement, in 1910, the International Conference of Working Women was held in Copenhagen. The event was organized with the leadership of a woman named Clara Zetkin, leader of the "Women's Office" of the German Social Democratic Party. It is at this conference that the idea of an "International Women's Day" to be celebrated every year in every country was first tabled. People from 17 counties actively participated.

In 1910, the great Copenhagen "International Women's Day" with the leadership of Clara Zetkin was ended with showing honor to millions of women from Denmark, German, and Switzerland for their demand on social equality, removal of discrimination and right for equal healthcare as compared to male as counterpart. Subsequently, after a couple of years, in 1913–14, when the people started campaign for peace to get relief from the distress consequence of World War, the Russian women observed their first International Women's Day on February 23 to bring awareness on women's equality in every sphere of life. But, at later stage the date for International Women's Day was changed to March 8. The date of March 8 was chosen because on this day Soviet Russia started protests for the right to vote which they were granted in 1917. In 1914, women across Europe held rallies to campaign against the war and express women's solidarity. In 1917, Russian women held strikes for "bread and peace" in response to the death of over 2 million Russian soldiers in World War I, and were granted right to vote. France was the only western nation in the world that did not give voting rights to women until 1944. In 1946, the French women started raising their voice for right to vote (Fig. 5.3). On October 27, 1946 it was de Gaulle's government that granted women right for vote.

In 1975, the United Nations for the first time celebrate International Women's Day but in the name of "United Nations Day for Women's Right and International

FIG. 5.3

Young French feminists demonstrate with posters that state "French Women want to vote" 1946.

(Source: Image Gallery).

Peace" on any day of the year by the member states, in accordance with their historical and national tradition.

Subsequently, International Women's Day has been celebrated annually, and offers new opportunities and innovative ideas for the well-being of women worldwide. In continuation, October 15 is the UN International Day of Rural Women, which is organized on an annual basis to promote the critical rule and contribution of rural women in agricultural development and the microeconomics of rural communities, which is ultimately linked with urban development.

5.2.2 Current global feature of international women's day

March 08 of every year has been declared a national holiday for International Women's Day in 27 countries around the world. On this day men acknowledge the women who are special in their lives with a bouquet of flowers as the insignia of deep gratitude and love. This short of practice helps in developing social inclusion which is important for a person's dignity, security, and opportunity to lead a better life. This day is also celebrated by many more countries through events such as rallies, conferences, and parades. It is the day for women from all cultures and social backgrounds to come together on a common platform to promote rights and raise awareness of women's issues.

To celebrate International Women's Day (IWD) is not mandatory for all over the countries around the world. However, IWD became a mainstream global holiday following its adoption by the United Nations in 1977. International Women's Day is commemorated in a variety of ways worldwide; it is a public holiday in several countries, and observed socially or locally in others.

5.2.2.1 China

In 1949, after the founding of the People's Republic of China, the government has recognized International Women's Day and made it an official holiday. China celebrates International Women's Day every year in several ways, including granting women a half day off work, discounts on women's products, and much more. The Chinese celebrate International Women's Day by showing a lot of respect and love. Girls from educational institutes celebrate 7th of March as "Girl's Day" instead of 8th March as IWD. This is mainly to bring cordial relation in between the students to keep peace and harmony. On this day men treat women with special surprises, gifts, cards, and flowers, and show their respect for women's rights (Fig. 5.4).

In this day, in order to bring social inclusion primary focus is given to women in the meeting organized for award ceremonies recognized for their contribution for nation's development.

5.2.2.2 Saudi Arabia

In 2017, Saudi Arabia started celebrating International Women's Day on February 01–03, as a 3-day event, instead of March 08. During this period both the royalty and common citizens come together to discuss women's rights by organizing social get-together and cultural programme. In recent years, Saudi Arabia has been heavily criticized for its record on gender equality and was ranked 134 out of 145 for

FIG. 5.4

Chinese women displaying pamphlets for women's rights on International Women's Day.

gender equality by the World Economic Forum. So, with the hope of elevate the status of women in the society, on the occasion of IWD, the Saudi Arabia honored women for their significance role in cultural activities, innovative achievement in the field of medicine, literature, and other social welfare activates (https://www.languageinsight.com).

5.2.2.3 Russia

Russians celebrate March 08 as a holiday for International Women's Day. The main theme for this is the fight of women from all over the world for their rights, full equality with men, democracy, and peace. This spring holiday is cordially celebrated by visiting kith and kin with exchange of thoughts on women's problem and possible solutions. During these meetings, the family members also share gift cards, flowers and chocolate, and other pleasant gifts with their mother, sisters, wives, grandmothers, and daughters. A flower bouquet is the most popular gift followed by candy, perfume, and cosmetics. This is mainly for developing social inclusion and harmony. Greater social inclusion means people are less likely to experience discrimination-based adversity.

5.2.2.4 Italy

In Italy, International Women's Day is known as *La Festa Della Donna.* On this day the male family members offer mimosa flowers to women as a gift of love and respect, similar to red roses on Valentine's Day (Fig. 5.5).

The yellow flower is chosen partly because it blooms in early March, and because they are often cheap. International Women's Day is not a public holiday in Italy, but on this day, women receive free access to museums across the country, as well as free medical check-ups from over 200 hospitals.

5.2.2.5 Poland

Following World War II, Poland's socialist governments campaigned for International Women's Day to promote the image of a woman as a respected worker, who supports

FIG. 5.5

Italian man gives a mimosa flower as a gift on International Women's Day.

FIG. 5.6

Tulip flower is given as a token of love and respects to women in Poland in International Women's Day.

her country. On that day, it was mandatory to celebrate in workplaces and schools. In this day women would receive carnations and other products such as towels or coffee. Today, the most popular gift is the tulip flower, which is presented by men irrespective of their position as boss, colleague, or friend (Fig. 5.6) to women. Tulip flower is also exchanged as a gift representing love and respect among family members. On International Women's Day, Poland celebrates the achievements of women and promoting maximum respect to women with valuable gifts. In large cities one can see young men with bunches of Tulips, handing them out to women on the street. By this short of practice everyone can participate fully in life and exercise their basic right in order to develop sustainable pattern of social structure and function.

5.2.2.6 Bulgaria

Since 1944, with the arrival of the socialist regime, Bulgaria has celebrated a national holiday for International Women's Day, but its popularity grew immensely in the 1960s. Currently, the holiday for International Women's Day is celebrated to appreciate all women, especially mothers. Women receive flowers, chocolates, and cards from their loved ones. In school, students give handmade cards to their female teachers, and create cards in class that can then be given to their mothers. Many companies buy red roses for each of their female employees. This is to express solidarity in social activities on the occasion of IWD, and to develop awareness on gender equality to elevate women in the society.

5.2.2.7 UK

In the UK, International Women's Day is celebrated with events such as workshops, talks, and demonstrations. The day is seen as an opportunity to discuss women's rights and raise awareness for the fight against gender discrimination.

5.2.2.8 India

In India, International Women's Day has not yet been declared a national holiday, but people actively participate in celebrating this occasion through marches on the street, and public meetings to raise awareness of women's rights and the need to remove inequality between men and women in all spheres of life. In India, in particular, meetings and public gatherings are organized in rural areas to explain the importance of wage equality between men and women working in the agricultural field.

5.2.3 Special theme for International Women's Day 2021

UN Women is the United Nations entity dedicated to empowerment of women, globally. A global champion for women and girls, UN Women was established to accelerate progress on meeting their need worldwide. Women keep updated touch with COVID-19 pandemic and looking after women's physical and mental health during COVID-19. UN Women declared the theme for International Women's Day 2021 (IWD 2021) as "Women in leadership: Achieving an equal future in a COVID-19 world." Currently, women are the main resource in the healthcare institutions for carrying out effective and inclusive COVID-19 responses, from the highest levels of decision-making to frontline service delivery. This theme is aligned with the theme of the 65th session (which took place from March 15–26, 2021) of the Commission on the Status of Women: "Women's full and effective participation and decision making in public life as well as the elimination of violence for achieving gender equality and the empowerment of all women and girls."

Through their efforts and input, women leaders and women's organizations are effective in leading the COVID-19 response and recovery efforts. Through this, it can be seen as never before that women bring different experiences, perspectives, and skills to the table, and make irreplaceable contributions to decision-making, policies, and laws that work in better ways in confronting the risks of the COVID-19 pandemic. Nevertheless, in spite of the huge contribution of women in the frontline workforce, still, there is disproportionate and inadequate representation of women in national and global COVID-19 policy spaces.

Overall, women form the majority of the health workforce that is stemming the tide of the COVID19 pandemic and responding to its health and brooder socioeconomic impacts, and those workforces are headed by women. For instant, the government of Denmark, Ethiopia, Finland, Germany, Iceland, New Zealand, and Slovakia have admire the highly sincere service and dedication of women healthcare work force for timely contribution of support in taking care of remediation measures in controlling COVID-19. Although women are serving on the frontlines against COVID-19, but the impact of the crisis on women is stark [3].

Keeping in view the dynamic and successful role of women in combatting the COVID-19 pandemic, it is high time to integrate women in a variety of roles, formulating policies and programs in all spheres and at all stages of pandemic response and recovery.

5.3 The suffering of women during COVID-19

The 25th Beijing Conference on September 4, 2020 set a path-breaking agenda for women's rights. The conference was a gathering of 30,000 activists representing 189 nations, who unanimously adopted the Beijing Declaration and Platform for Action. The main theme of agenda is a vision of equal rights, freedom, and opportunity for women everywhere, no matter what their circumstances, that continues to shape gender equality and women's movements worldwide [1–4].

The Beijing Platform for Action sets the objective of achieving a balance between women and men in national decision-making positions. However, there are many countries that still ignore the balance factor between women and men in distributing power at government level. According to the Inter-Parliamentary Union (IPU), women represented, on average, 21.4% of all parliamentarians in 187 countries, as of 2013. A survey by IPU shows that since the Beijing Conference attitudes and awareness have improved, but this has yet to lead to remarkable changes in implementing equality in public and political life.

Furthermore, with the rapid spread of COVID-19, whatever little gains have been made in achieving women's equality are at risk of being rolled back. The life-threatening pandemic magnifies the preexisting inequalities, exposing vulnerabilities in social, political, and economic systems.

Both women and girls are victims of COVID-19 in all spheres of life, covering health, wealth, economy, and social protection [5–8]. The earning potential of women and girls has, in many cases, succumbed to the financial crisis, with temporary employees having to leave their jobs. Girls, especially those from rural areas of developing countries, are heavily affected by the secondary impacts of the outbreak. The financial burden on families due to the outbreak can put children, and particularly girls, at greater risk of exploitation as child labor and of gender-based violence.

Due to "lockdowns" and restriction in movement because of COVID-19, women, especially in rural communalities, have experienced social isolation and faced gender-based violence, mainly due to financial crises in the family [9–12].

Casualty of millions of people in COVID-19 has challenged the world to serve the people without any compromise to bring normal life again. The global economic damage is such that to gain sustainable economic conditions appears out of reach. The microeconomics of rural communities in developing countries have been so reduced that to recover from this great crisis is a Herculean task. Women will be the hardest hit by this pandemic, but they will also be the backbone of recovery in communities. Therefore, it is necessary to empower rural women to lead agricultural development. In addition, in this critical situation, maximum attention should be paid to rural women overcome with the harm caused by the COVID-19 pandemic. In spite of the horrifying situation of COVID-19 it is highest challenge before us to bring proper remediation measures to give relief to the COVID-19 contaminated people with special attention on rural women, girls, and children. The following are some of the important facts and figures on COVID-19 affected areas around the world.

5.3.1 Diversified risks from COVID-19

The joint statement by the International Labour Organization (ILO), International Fund for Agricultural Development (IFAD), Food and Agriculture Organization (FAO), and World Health Organization (WHO) says that the COVID-19 has caused loss of millions of human life all over the world with the worst effect on public health, food security, and normal human activities. The economic and social disruption caused by the pandemic is devastating, leading to 10 million people being at risk of extreme poverty, undernourishment, and mental and physical illness.

About 3.3 billion members of the global workforce are at risk of losing their livelihood. In particular, workers from informal economy are vulnerable, due to inadequate access to production resources. The lack of earning opportunity makes maintenance of quality healthcare a challenging issue when even feeding themselves and their families is difficult. Poor farmers are confronted with the problem of border closers, trade restrictions, and confinement measures affecting access to markets, including for buying inputs and selling their products. This has resulted in discontinuity of the food supply chain and risks to food security. The pandemic has resulted in loss of jobs for people, particularly women, who, without daily wages, are unable to manage their livelihoods and children's healthcare. As food producers lose their jobs under extreme financial conditions, the food security and nutrition of millions of women and men are under threat, with those in low-income countries, particularly the most marginalized populations, which include small-scale farmers and indigenous peoples, being hardest hit.

Currently, many survey reports are in the process of emerging on the negative impacts of COVID-19 on health, especially of people residing in rural areas and least developed countries. The COVID-19 dashboard by the Centre for Systems Science and Engineering at John Hopkins University, 2020, disclosed that currently, there are 23 million confirmed cases with 800,000 deaths in COVID-19 are available globally. The pandemic has taken into its grip many women and girls who have limited access to healthcare services due to the unequal allocation of resources and priorities. The largest gender gap in COVID-19 was noticed among those aged 60–70.

5.3.2 Economic crisis and unemployment

The impact of COVID-19 on the global economy has been to cause great loss, and it seems it will be a Herculean task to return to normal. The COVID-19 pandemic has resulted in problems of unemployment in many countries [13]. It has been estimated that about 2.2 to 2.8 billion youth are unemployed due to the partial or total lockdown in COVID-19 pandemic [14–16]. This sudden mass loss of employment not only affects the individuals concerned, but also paralyzes the resources for many social support systems [17]. Many youths, including women and girls, are suffering from mental depression [18]. The health costs of COVID-19 not only affect the individual, but also damage family financial conditions [7–10,19–22]. Plenty of information is available in the literature on the various aspects of depression [23–29].

The International Labour Organization (ILO) has estimated that full or partial lockdown measures now affect about 2.7 billion workers, representing about 81% of the world's workforce [30], while the International Monetary Fund (IMF) projected a significant contraction of global output in 2020 [31].

Both developed and developing countries are equally affected by the unemployment crisis due to the sudden outbreak of COVID-19. For example, the present unemployed record in India is about 27.1%, according to the Centre for Monitoring the Indian Economy (CMIE). The recent data shows India's unemployment figures are four times that of the United States. India does not release official jobs data, but CMIE data is widely accepted. This unemployment is mainly due to lockdowns to curb COVID-19. A closer look at CMIE's data shows the devastating effect the lockdown has had on India's organized economy (Fig. 5.7).

In India, of the 122 million who have lost their jobs, 91.3 million were small traders and laborers. In addition, a fairly significant number of salaried workers (17.8 million) and self-employed (18.2 million) have also lost work. Meanwhile, the government has lifted lockdown prohibition in areas which have recorded lower numbers of infections, while strict curfews are still in place in districts that have heavy positive numbers of COVID-19. Although zoning is a good practice, in the long run it results in adverse effects. In addition, the government has started a supply chain of a variety of goods required for daily life. The lockdown is slated to end on 17 May 2021, but some states have extended it further, with no clear indication as to when the country as a whole will emerge from lockdown.

It has been predicted that women's economic and productive lives will be affected disproportionately and differently. Women earn less and also save less. They have less access to social protections as they have less access to inherited family property.

FIG. 5.7

Small businesses (shopping market, India), which have been hit hardest by the lockdown.

Therefore, it is obvious that women have the potential to bear economic shocks less than that of men. As women manage the entire household work, their part time or full time work is disproportionately affected by the reduction in earning due to COVID-19.

The women in developing countries suffer in the worst way due to COVID-19. Under the informal economy, with limited protection, the majority of women feel helpless due to the scarcity of resources under pandemic disease conditions. To earn a living, the female workforce often depends on public spaces and social interactions. The corona virus showed that quarantines can significantly reduce women's economic and livelihood activities, increasing poverty rates, and exacerbating food insecurity. It is currently projected that the impact of the COVID-19 global recession will result in a prolonged financial crisis and reduced labor force participation, with compounded impact for women already living in poverty.

5.3.2.1 Economic recovery measures

The present and future action plan on risk management of the COVID-19 pandemic should be targeted to build more equal, inclusive, and sustainable economies at micro- and macroeconomic level in rural communities and at the national level, respectively. This is the most comprehensive challenge of the coronavirus pandemic for risk management, based on social policies and rural economics. On March 31, 2020, 105 countries resolved fiscal response packages equivalent to a total of US$ 4.8 trillion [32]. By April 03, 106 countries had adopted social protection and jobs programs in response to the pandemic [33]. Social assistance (noncontributory transfer) is the most widely used package, followed by social insurance and supply-labor market interventions. Cash-transfer programs are the most widely used social assistance intervention. In sectors largely populated by working women, where supply chains have been disturbed, there should be adequate provision to access credit, loans, and grants so as to retain the female workforce.

Besides these measures, the whole range of economic policies for both immediate response and long-term recovery should be planed and implemented to facilitate women to continue earning a living. In addition, barriers that restrict the full involvement of women in earning activities, such as unequal wages, should be removed. Finance should be provided for women entrepreneurs through simple official procedures without any harassment. There should be easy mechanisms and technical guidelines for self-employed women linked with public and private spheres.

In order to recover from the economic crises caused by the COVID-19 pandemic, it is now high time for narrowing gender-based educational gaps, and ensuring women's participation in the formal labor market will play a significant role in promoting sustainable economic growth. In fact, social protection systems effectively do not cover women in the workforce in many cases. For example, in South Asia, over 80% of women in nonagricultural jobs are in informal employment; in sub-Saharan Africa this figure is 74%, and in Latin America and the Caribbean, 54% of women in nonagricultural jobs are in informal employment. It is high time that governments—in particular, of developing countries—should initiate the provision of access to benefits such as health insurance, paid sick and maternity leave, pensions, and unemployment benefits in ways that reach beyond formal employment and are accessible to women in all spheres of work.

5.3.3 Healthcare

During the COVID-19 pandemic, it has been more problematic to provide access to healthcare, especially in rural areas of developing countries. This is catalyzed by multiple or intersecting inequalities, such as ethnicity, socioeconomic status, disability, age, race, geographic location, and restriction of critical health services and information to women. Health services for men and women are equally important, but due to some particular features in women, like maternal health, they need special attention. In rural areas, the insurance coverage for women is very poor: most rural women are excluded from accessing insurance at an affordable cost. Social exclusion due to restrictive social culture and roles and gender stereotypes can also limit women's ability to access health services. All of this has particular impacts during a widespread health crisis like the COVID-19 pandemic.

Women form the frontline of health workers, especially as nurses, midwives, and community health workers. They also represent the majority of health service-staff, such as cleaners, laundry workers, and caterers, and as such, are more likely to be contaminated with contagious diseases. In some areas, women have less access to personal protective equipment. For example, personal protective aprons used in attending COVID-19 patients are not designed as per the requirements of female staff members. Despite all of these problems existing for women health workers, they are largely not included in national or global decision-making on the response to COVID-19.

The impact of COVID-19 on sexual and reproductive health of women seems to be alarming. Maternal healthcare and gender-based violence related services should be treated as a priority. Even a small diversion of attention from these areas may result in exacerbated maternal mortality and morbidity, and increased rates of adolescent pregnancy, HIV, and sexually transmitted diseases. For example, in Latin America, it is estimated that an additional 18 million women will lose regular access to modern contraceptives, given the current context of the COVID-19 pandemic [34].

It is a significant challenge to know how to respond to the health impacts of COVID-19. It is critical for all public health preparedness and response plans to COVID-19 to consider both the direct and indirect health impacts on women and girls. First of all, it is most important to convey the public healthcare message to women and girls on how to tackle the health-related issues resulting from COVID-19. With limited access to education, care must be taken to pass on the message to the public through involvement in cultural activities and community development. In addition, it is important to distribute materials related to COVID-19 prevention to women and girls residing in rural areas and those that are in refugee and settlement camps.

Special provision should be made for female health workers and nurses that are serving on the front line. They should be provided with personal protective equipment (PPE), safety approve mouth cap (mask) which block transmission of bacteria or viruses. The protective equipment should be the appropriate size for women. In most cases, women are provided with protective equipment designed for men, and this sort of discrimination causes a lot of inconvenience. It is essential to provide women frontline workers and those serving in quarantined areas with essential hygiene and sanitation items (e.g., sanitary pads, soap, hand sanitizers, etc.). Frequently, female health workers face public violence and abuse against them. Therefore, it is necessary to take decisive measures to prevent them being exposed to such behavior while serving patients.

It is necessary to give priority to healthcare services for older women, gender-based violence survivors, those needing emergency obstetrics, and newborns in areas with prevailing COVID-19 infections, through healthcare centers. Necessary infection control measures should be in place. HIV treatment access needs to be maintained with no interruption, particularly, but not exclusively, in terms of prevention of mother to child transmission of HIV.

Many developed countries like the United States, with the help of the World Health Organization (WHO), are trying to take into account the gender dimension inherent in the coronavirus pandemic by strengthening their health systems and other health services. In this context, the UN supports the state governments of countries with weak public health and social support systems. While serving the global population, the UN keeps especial attention on rural women and girls, including higher risks groups such as pregnant women, people living with HIV, and persons with disabilities, in protecting them from the risk of COVID-19 contamination. Therefore, the UN has been putting maximum effort, with governments and partners, into ensuring real functional activities to reduce the risk of sexual and reproductive health problems, and protect the rights of women and girls.

The COVID-19 pandemic has increased the struggle of shouldering the healthcare burden on families and communities, which has been increasing by the day. In hospital, it has become regular practice to discharge COVID-19 patients as soon as possible to make space for others, but these patients still require care and assistance at home. Postclinical care of COVID-19 is of equal importance to hospital treatment. It is often women that have to take dual responsibility for preexisting chronic patients at home and COVID-19 patients discharged from hospital. Women are at the forefront of the COVID-19 response, as the default unpaid family caregivers and the majority of unpaid or poorly paid community health workers. Lockdowns and the closing of schools have added strain and increased the demands on women and girls at home. As stated by UNESCO, 1.52 billion students (87%) and over 60 million teachers are laying idle at home. Due to reduction in formal and informal childcare, the demand for unpaid childcare provision is falling heavily on women, not only because of the prevailing system of the workforce, but also because of social norms.

5.4 Women's unpaid labor

Commonly, the term unpaid labor is used to refer to work without any remuneration (noncompulsory work performed for others without pay). Own-use production work

FIG. 5.8

Unpaid domestic labor and the invisibilization of women's work.

refers to activities performed to produce goods or provide services intended for final use by the producer, their household, and or family (Fig. 5.8).

In developing counties, it is mainly women that take responsibility for unpaid care work in the family. Women do about 75% of total unpaid work, globally; or in other words, three times that done by men [35].

Due to limited access to healthcare services and issues of unequal allocation of resources and priority, women and girls have been adversely affected by COVID-19. Women with the age of 60–74 are more vulnerable to COVID-19 as compared to men with the same age group. The lockdowns due to COVID-19 have not only restricted the movement of older people, but also reduced the provision of care provided by older women.

Lockdown measures like closure of national borders have overturned the global economy, leaving behind millions of workers worldwide. As reported by the International Labour Organization, 2020, about 14% of working hours, an equivalent of 400 million fulltime jobs, were lost during the second phase of the coronavirus pandemic [36] globally. This situation is worse than the job losses during the global economic crisis of 2008–09. The COVID-19 effects have caused most devastation in South Asia and Sub-Saharan Africa [37] women are more significantly affected. The closing of schools in more than 190 countries has affected about 1.6 billion students [38], interrupted the internet, and disabled digital technology. The pandemic has caused severe damage to food supply chains, placing about 265 million people in food crisis.

In the second phase of the coronavirus pandemic, unpaid care work performed by women was totally ignored. This is mainly due to inadequate financial budget of government, mostly in developing countries [37,39]. More than 510 million workers, including women and girls, from hard-hit sectors that include retail, hospitality, food service and manufacturing, especially the garment industry, [36] have become victims of lockdowns. About 96 million or 70.4% of the total workforce in the health and social work sector working on the frontline of COVID-19 have lost their jobs [40]. COVID-19 has caused tremendous effects on people belonging to minority ethnic communities [41,42], which is ultimately responsible for damage to socio-economic conditions, the livelihood status of families and communities, and unequal employment opportunities in the essential healthcare sector.

Over the last decade, women's unpaid work has become a highly importnat topic discussed all over the world, and economists have been trying to incorporate women's unpaid work into the mainstream of economic policy [43–45].

Unpaid work plays a significant role in economic activity and is essential for the well-being of individuals, households, communities, and at the national level. However, in most of the developing countries the government ignore the importance of unpaid work for economic production of goods for self-consumption or services enjoyed by others in the household.

Globally, outbreaks of disease worsen the issue of gender inequality. Prior to the coronavirus pandemic, women and girls were responsible for unpaid work totaling about 12.5 billion hours, globally [46]. The sharp gender disparity is clearly visible in South Asia. Women in India spend 10 times more time on care work than men, both in urban and rural areas, whereas in Bangladesh, women spend about three times the time that men do. The unpaid labor of women, while fueling economic growth, has deprived women and girls of time and resources for education, skill development, or for gainful employment. Unpaid care work, a driver of inequality, places women under tremendous mental tension, risking depression, with insecure incomes and lack of social security. The COVID-19 pandemic has intensified the load on care systems, already depleted and unfair, falling mostly on the shoulders of women.

In developing countries like India, the rate of payment of women is lower than man's and has been in declining trend over the last decade, maybe due to the preference of women often being to do domestic work rather than low labor cost public work [47]. In India only 22% of women are involved in the workforce, and out of them, 70% are associated with informal farm activities with minimal daily wages without any protection guarantee [48]. In India, compared to urban women, rural women do more unpaid care work (Fig. 5.9).

Generally, in most of the developing countries, women's unpaid work is unnoticed and unrecognized. The main reason for poor involvement in the workforce of rural women is that their socioeconomic position in the society is to carry out unpaid household activities in the family in the form of cooking, cleaning, fetching water and fire wood, and taking care of family members [49,50]. Women carry out at least two and a half times more unpaid household care than men. In addition, rural women lack opportunities to access adequate public provisioning in critical sectors such as energy, health, water and sanitation, food security, and livelihood [51]. Globally, per day men spend 83 min in unpaid domestic work as compared to 265 min for women [52]. Women's unpaid care work is equivalent to 10%–30% percent of gross domestic product (GDP), which can contribute more to the economy than the manufacturing, commerce, or transport sectors [53].

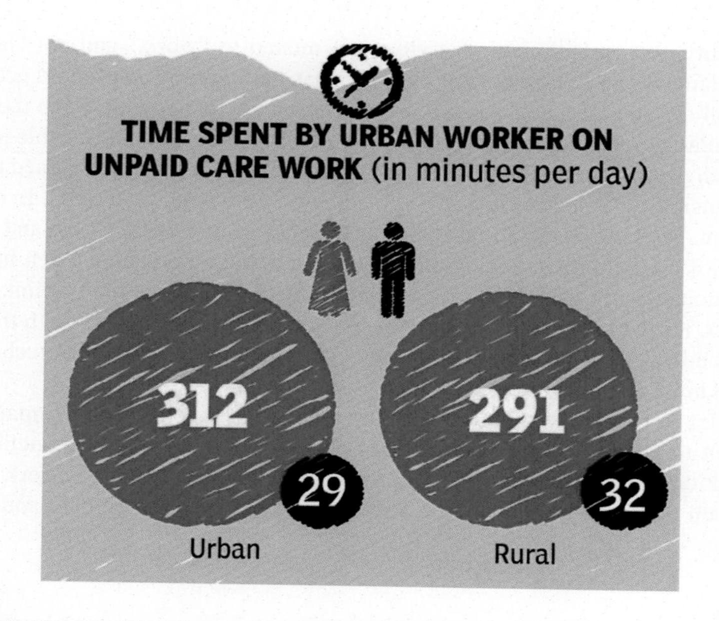

FIG. 5.9

In India rural women spend more time in unpaid care work as compared to urban women.

The unpaid care provided by households is the most important part of care, as it keeps families together and nurtures human and social values. Unpaid care services play an important role, particularly in countries where basic infrastructure and public services are weak.

5.4.1 Factuality of unpaid work

Economists have had reservations about accepting unpaid labor as part of the mainstream economy under different economic theories. Both the classical and neoclassical economists consider unpaid work as outside the production purview rather than accepting it as an economic good or market good. Furthermore, while preparing for national annual budgets, unpaid work value is not considered in budgetary provision. In this way, women's unpaid work is seen as part of a housewife's life rather than a part of the economy. However, in 1960s, this interpretation was counterattacked by some feminist economists by including women's domestic labor into the domain of economics and analyzing it as a form of work comparable to paid work [54–60]. This study was conducted to understand the factual value of domestic work being carried out by women as unpaid labor, and to understand that many women are not present in the labor market, not because of their personal choice, but due to their economic disadvantage and low opportunity cost. The overall target was to bring into the limelight the factuality of unpaid work of women in all spheres of activities.

In reality, women's dometic activities are related to "work," and not "leisure." So, it is claimed that the concept of "work" means all types of household activities. The sexual division of labor is based on socioeconomic patterns between the sexes. Gender polarization, gender freedom, and gender integration are all possible patterns of sexual division of labor [61]. In gender polarization, paid work is assigned to men, and responsibility for unpaid household work is given to women [62–65]. In the second stream, gender freedom, opportunity is given to women both in paid and unpaid work [6,66]. In the third stream, gender integration, both men and women integrate paid and unpaid work [67]; in this integrative form, these stresses are interlinked with each other. Furthermore, the status of women's unpaid work can vary with the passing of time because it is influenced by race, class, ethnicity, transfer of technology, and the state of capitalism [68].

In order to overcome with the various issues related to women's unpaid work it was felt to tackle this problem in three phases, i.e., recognition, reduction, and redistribution under the caption of "triple R" in order to integrate unpaid work into the mainstream economy by reducing it and by reorganizing between paid and unpaid work (Fig. 5.10) [69].

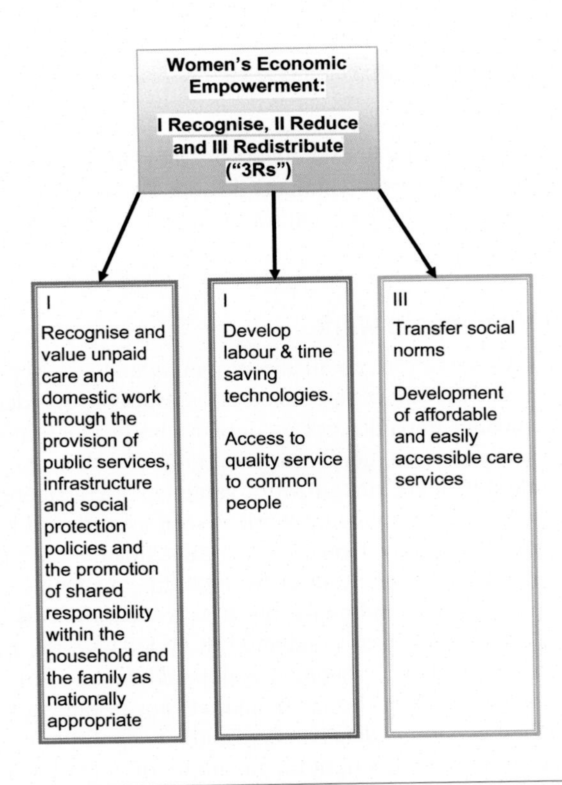

FIG. 5.10

The 3Rs approach: three interconnected dimensions to address unpaid care work.

5.4.2 Women's domestic activity

Sustainable Development Goals Target 5.4 places emphasis on women's unpaid labor in terms of social protection, public service provision and infrastructure, and the promotion of shared responsibility (Fig. 5.11).

By redistributing the tasks of unpaid labor among men and women, women can take the time to empower themselves for earning. By performing unpaid work, women subsidize the market and also reduce the burden of the state [51]. While analyzing the scenario of global economy the neoliberal macroeconomic policies have been formulated ignoring unpaid work into account, and mentioning its negative impact on the global economy [60,70].

In general, unpaid work is not a matter of choice for women, but is rather a compulsory assignment by society as a regular practice at home. These sorts of practices restrict women's ability to participate in the labor market or other work [71,72]. In addition, demographic factors, mainly fertility rates and family structure of women, play a significant role in the status of women as unpaid workers [73,74]. Furthermore, basic infrastructure for health and education, safe water, sanitation, energy for lighting and fuel, transport, and childcare also have a strong influence on the time spent for unpaid work. In developing countries, failure of the state to provide alternatives for care and domestic assistance increases the burden of unpaid work [75,76]. In rural areas of developing countries, the unpaid women are not accessed properly while enrolling their name for self-employ, casual part-time work, or any other seasonal work without any kind of security [77,78].

5.4.3 Women's unpaid work during COVID-19

The coronavirus outbreak at a global level starkly highlighted how, in reality, the world's formal economies and the maintenance of our daily lives are firmly based on the unnoticed and unpaid work of women and girls, both in the family and the community. The COVID-19 pandemic has intensified the care needs

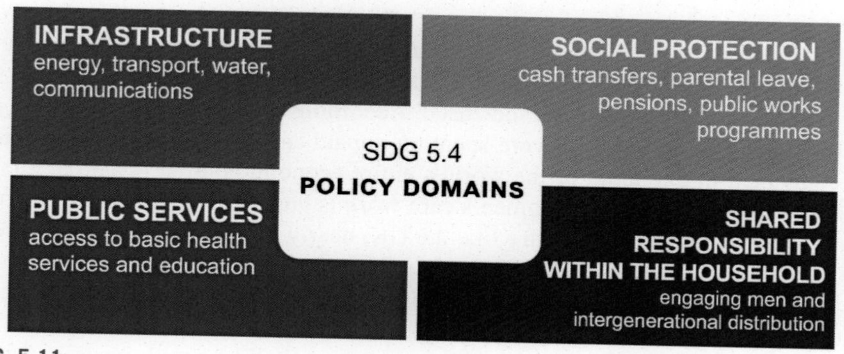

FIG. 5.11

Various aspects of the Sustainable Development Goal SDG5.4.

for older persons, and school-age children. Caring for ill family members, and emergency health services being overwhelmed, has increased demand for care work during COVID-19. Therefore, it is obvious that the unpaid care economy is critical to the COVID-19 response. There are gross imbalances in the gender distribution of unpaid care work. Before COVID-19 became a universal pandemic, women carried out times as much unpaid care and domestic work as men. This unseen economy has real impacts on the formal economy and women's lives.

The global outbreak of COVID-19 pandemic and the present scenario of unequal gender division of labor has worsened the risks to healthcare systems, particularly in situations where health systems are overloaded due to COVID-19 lockdowns with the closing of offices, industries, commercial centers, and schools and other educational institutes. Under this critical situation, women's unpaid labor would be immensely helpful to provide timely health care service to COVID-19 patients. However, it could only be possible with the guarantee of full protection of unpaid women from violence and abuse caused by the patients and public within and around the immediate surroundings of hospitals and temporary healthcare shelters.

The health risk crisis caused due to the coronavirus pandemic has, in an unprecedented way, challenged the world with the critical role of care, which can be mainly performed by women, both as frontline healthcare workers and informal care providers in their families. So, women are the overall source of unseen power and caretakers who are able to roll back to our preexisting normal life in the post-COVID era. So, women unpaid health care force is one of the major sources to be deputed for taking care of COVID-19 patients, and their timely recovery to normal life. So, it is high time to avail the service of women's unpaid care work both in hospital and to attend house quarantine COVID-19 patients.

It is a fact that the COVID-19 pandemic has hit everyone hard, but the impact on health risk is unevenly spread. The load of unpaid care work has increased both for men and women, but the unpaid care workload more on women than men. This is because of women have to take care of quarantine COVID-19 patient in addition to taking care of entire household work like cooking, taking care of children education, looking after elderly members of the family, and other managing agricultural activities in the crop field. After the outbreak of COVID-19, the unpaid work load has been increased. According to an ILO survey report, in 2018, about 16 billion hours were spent on unpaid caring every day. This would be equivalent to one tenth of the world's entire economic output if paid at a fair rate. In India, about 66% of women's care work is unpaid as opposed to 12% for men [79]. In a developed country like the UK, nearly three-fifths of unpaid cares are women, and they are more likely than men to be caring for someone living in another household [80]. In the United States, disproportionately, it is black, Latino (of Latin America origin), and immigrant women that are employed as care workers with low wages and without any benefits like health insurance and retirement security [81].

Women's unpaid contributions to healthcare is equivalent to 2.35% of global GDP or about US$ 1.488 trillion [82]. It is predicted that this value will increase to 9% as women play vital role in the global healthcare as nurses, midwives, community health workers, and doctors.

During the COVID-19 pandemic, paid healthcare workers are mostly female, not including those women and girls serving at home unpaid care workers, taking care of elderly persons, children, and chronic patients [83]. Most countries in the world are facing shortages of unpaid care workers due to lack of protection and security. The lockdown of educational institutes, offices, and other working places has also resulted in closing of daycare centers [84].

Closures of schools due to lockdown has increased the workload of child care on women and girls. UNESCO reports that about 1.5 billion students and 63 million primary teachers from 188 countries are confined to their homes [85]. Consequently, at home the additional burden of women and girls to take educational care of children is increased. In many countries, such as India, people have lost their jobs in the formal sector [86]. Many women are forced to work for low daily wages, which often causes mental depression [87]. About 80% of the world's domestic workers are women [88]. Prior to COVID-19, many domestic workers use to migrate around Southeast Asia between the Philippines, Indonesia, Hong Kong, and Singapore. By such practices women sent $300 billion home every year, half of the total global remittances. However, due to lockdown, migrant women are losing jobs [89].

As caregivers, women and girls face higher risks of infection from COVID-19. Care work involves personal interaction, where social distancing is difficult to practice. Globally, 70% of the poorly paid healthcare workers are women, without any security guarantee. In China's Hubei province, where the virus originated, women are 90% of the health workforce [90].

The Association for Sanitation and Health Activities (ASHA) is a nongovernmental organization (NGO) based in New Delhi. It works at a grassroots level to reach out to the most vulnerable and marginalized communities to empower them for their self-reliance and contribute towards their social and economic uplift. However, the ASHA workers have no fixed salary nor are they supplied with any personal protective equipment or training, and access to portable instrument for testing COVID-19 patients. It is also mandatory that each worker has to visit 25 homes per day to screen suspected patients in both rural and urban areas. Mostly ASHA voluntary organization prefer female member rather than male because of to undertake family related assignment like counsel women, families and adolescents birth like issues [91].

Rural and urban women alike are often insecure even at home as they have to take on full responsibility for COVID-19 patients discharged from hospital and also those that are in quarantine at home. While women are at the frontlines of the coronavirus pandemic, their own health keeps being pushed to the lowest priority. Women and girls, by virtue of gender inequality in every sphere of the health system, including economy, security, and other activities, are badly trapped by the risks of the pandemic.

Across every sphere, from health to the economy, security to social protection, the impacts of COVID-19 are exacerbated for women and girls simply by virtue of their sex. The following are a few salient features that have emerged during the COVID-19 pandemic:

- Women and girls who have limited access to earning, with job insecurity, feel unsafe due to the accumulated economic impact of the coronavirus pandemic.
- In the earlier stages of the pandemic, more men than women became ill, which resulted in increasing responsibility for women by bearing additional workloads including sexual and reproductive health services.
- Lockdown due to COVID-19 has resulted in school age children being confined at home, with it overwhelmingly being women and girls responsible for monitoring their health.
- During COVID-19, gender-based violence has increased due to restrictions in movement and social isolation measures with the implementation of curfews.

All these impacts are further magnified due to social exclusion, conflict, and ethnicity at community level. Therefore, it is necessary to amend policy at government level to explore the participation of women and girls in decision-making bodies to remove the barriers faced by women and girls in rural and remote areas. This is in the interest of not only women and girls but also boys and men. Although women are the hardest hit by the coronavirus pandemic, they are also the backbone of recovery in communities. To achieve this, the following action measures can immensely be helpful:

1. There should be female representative including frontline women workers, in all types of COVID-19 risk management bodies to take care of adequate provision of wages (economy planning), emergencies response (with necessary healthcare aids), protection from violence at working sites (hospitals and temporary shelters), and supply of personal protection equipment for frontline workers.
2. To consult at government level for driving transformative change for equality by addressing the care economy in the formal and informal sectors, like teachers and nurses, both in government and other sectors.
3. On the basis of the international understanding of gender equality, it is necessary to frame fiscal stimulus packages and social assistance programs to achieve greater equality, opportunities, and social protection.

5.4.4 Gender dynamics in unpaid care work

Currently, in the COVID-19 pandemic, one of the most challenging tasks is to regain human wellbeing in societies through unpaid or paid care work based on the female work force. Unpaid care work is one of the most important pillars supporting economic activity, both on a daily and a general basis [92].

The International Labour Organization (ILO), in 2018, estimated data on unpaid care work from 53 countries and revealed that the value of unpaid care is equivalent to 9% of the total global GDP amounting US$ 11 trillion [93] of purchasing power. At present, all over the world, women do more unpaid care work than men throughout over the course of their lives. On average, globally, women perform 76.4% (three quarters) of the total amount of unpaid care work. When it is expressed on the time basis, women's unpaid work represents 3.2 times more hours as compared to men [94], which is equivalent to a total of 201 working days for women and 63 working days for men, without any remuneration. Globally, the share of the population aged 65 years or over increased from 6% in 1990 to 9% in 2019. That proportion is projected to rise further to 16% by 2050, so that one in six people in the world will be aged 65 years or over. Therefore, due to population aging, large numbers of elderly women in the population can act as unpaid care providers [95,96]. They spend on taking care of house hold work like cooking, looking after the children, and other related work for family wellbeing. But still, the progress in reducing the unpaid care gap has been slow.

This is because of nonexistence of any mandatory acts to have control over gender inequalities at government level, and to pay special attention on removing disparities in assigning unpaid work to women and men for welfare of the society. Over a 15-year time span, across 25 countries with comparable data, women's unpaid care work decreased by only 10 min, whereas men's unpaid care work increased by only 13 min, with women continuing to spend disproportionately more time on unpaid care work than men.

Complete and partial lockdown has restricted the movement of male members of the family, and allowed them to share in unpaid household work such as shopping, transportation, and maintenance of the house [97]. This sort of unpaid care work during the COVID-19 pandemic has mostly been Norway, Mexico, and New Zealand [94], and male members of the family have started availing themselves of parental leave, as practiced earlier in some countries [98–100].

Women typically spend disproportionately more time on unpaid care work than men. So, UN women reports the deep rooted facts on unpaid care work mainly resulted due to extremely low standard of livelihood and poor-economy condition with low educational background, and disproportion share of unpaid care work (ILO, 2018).

Due to inadequate infrastructure such as water and electricity, and lack of provision of transport, rural women are left with little time for any outside work. Due to poor education, rural women spend more time in unpaid care work as compared to educated urban women who spend little time on unpaid care wok. Gender gaps across different races and ethnicities play a significant role in unpaid car work. In the USA among Hispanic and Asian couples, women spend more time in unpaid care work as compared to male family members as counter partners [101–103].

One of the main reasons for unpaid care work, particularly in rural area, is the status of racial- and ethnic-based communities suffering from extreme poverty and managing their larger families with multiple generations, with less access to childcare and services [104].

All over the world, the amount of time women devote to unpaid care work depends on the presence of children in a household. This results in a so-called "motherhood penalty." This has been intensified by the coronavirus pandemic. The term motherhood penalty is often is used by sociologists, who argue that in the workplace, working mothers confront biological- and cultural-based disadvantages in receiving equal pay, perceived competence, and benefits, relative to childless women. The pay gap between nonmothers and mothers is a serious problem effecting the livelihoods of mothers. During the coronavirus pandemic, mothers face the problem of worse working conditions and job insecurity than nonmothers. These effects are not only seen in highly developed countries like the United States, but are also common in many other countries like Japan, South Korea, Netherlands, Poland, and Australia. The penalty has not shown any signs of declining over time [105].

5.5 The COVID-19 workplace and women's leadership

As explained earlier, COVID-19 spreads primarily through respiratory exhaled droplets or contact with contaminated surfaces. Exposure can occur at the workplace, while traveling to and from work, or during work-related travel to an area with local community transmission. Schools, educational institutes, industries, and electronic-communication network hubs are the main platforms for COVID-19 transmission among employers and visitors. Women who are at helm of institutions can show leadership in mitigating health risk issues caused due to the pandemic. Following action plan can be helpful to understand the significance of women role related to unpaid health care, and can be work out in different work place for counseling healthcare measures as per the guide line of WHO or UNICEF.

5.5.1 COVID-19-prone workplaces and control strategies

The coronavirus pandemic has forcefully changed day-to-day life in unprecedented ways. The only option for all sections of society, including employers and employees, is to mitigate the health risks by mutual understanding and efforts with the spirit of humanity.

When a person is contaminated with COVID-19, he or she exhales from mouth or nose droplets that can infect another's epithelial cells of the nose and mouth, which can then transmit through the blood stream to the lungs, causing serious breathing problems. Therefore, WHO advice is to maintain social distance outside the home and use a face covering, as recommended by healthcare providers. Exhaled droplets from a COVID-19-infected person can precipitate on nearby surfaces or objects. So, by touching such surfaces, the coronavirus passes into the human body through eyes, nose, and mouth. Most persons infected with COVID-19 experience mild symptoms and recover. But if care is not taken at the initial stage of infection, serious health problems may cause a threat to life. Elderly people are more prone to COVID-19 then the younger generation.

The World Health Organization suggests:

i) simple ways to prevent the spread of COVID-19 in any workplace;
ii) how to manage COVID-19 risks when organizing meetings and other events;
iii) while traveling how to take prevention measures; and
iv) how to take preventive measures for healthcare providers to maintain their workplaces free from contamination.

Keeping in mind the aforesaid suggestions, WHO objectives include: (a) simple ways to prevent the spread of COVID-19 in a specific workplace, (b) how to manage COVID-19 risks when organizing meetings and events, (c) things to consider when employees travel, and (d) getting workplace ready in case COVID-19 arrives in a community.

Meanwhile, the WHO is providing guidelines and advice and bringing awareness among people by providing up-to-date information on COVID-19 and how employers can protect their employees at work. The WHO, with the International Labour Organization, provides information on the process of transmission of pandemic disease in general in workplaces, and recommends preventive measures to mitigating risk of COVID-19 at various types of workplace. The prevention of COVID-19 in workplaces should cover all aspects of health, including physical and mental health, safety and well-being of workers, protection from other occupational hazards in the operation, and closures and reopening of workplaces. The joint WHO/ILO policies briefly explain the factuality of transmission of COVID-19 all around the workplace and how to tackle the situation within the framework of specific time and affordable budgetary provision.

Since the outbreak of the coronavirus pandemic in early 2020, UNICEF and partners have supplied critical health care equipment and also provided financial support to 153 developing countries and territories to take immediate measures to save millions of lives from COVID-19, and deal with other related problems like nutrition; education; child protection; clean water, sanitation, and hygiene (WASH); and gender-based violence and social protection services during the coronavirus pandemic.

To combat the impact of COVID-19 on socioeconomic deterioration, UNICEF has supported adaptations to service delivery systems to minimize delay, maintain continuity, and enable equitable access. In 2020, the WHO and UNICEF jointly trained 3.3 million healthcare workers to provide services for reducing the impact of COVID-19; 1.8 million health workers benefited from personal protective equipment; 73.7 million people received WASH supplies; and 93 countries received 15,000 oxygen concentrators and other innovative devices that help people with COVID-19 breathe. In addition, UNICEF intervenes with activities of community groups, health care providers, and other nongovernment organization to reach 3 billion people to protect them from the risk of COVID-19, globally.

5.5.2 Joint mission of the WHO and China

In January 2020, the WHO declared the first outbreak of a new coronavirus disease in Hubei Province, China, to be a public Health Emergency of international concern, from where high-risk coronavirus has spread to other countries around the world. In order to combat the healthcare risk, the WHO and China jointly started a mission on next steps in response to the ongoing outbreak of novel coronavirus disease, COVID-19. The Joint Mission consisted of 25 members, including national and international experts from China, Germany, Japan, Korea, Nigeria, Russia, Singapore, the United States of America, and the WHO. The Joint Mission was headed by Dr. Bruce Aylward of the WHO and Dr. Wannian Liang of the People's Republic of China. The major targets of the Joint Mission were as follows:

- To know meticulously the root causes of the outbreak of COVID-19 in China and its effect at community level in China.
- To generate recommendations for adjusting COVID-19 containment and response measures in China and at the global level.
- To start a collaborative program of work, research, and development to mitigate the healthcare risks, and to update the knowledge and response for the public, worldwide.

The Joint Mission was implemented over a 9-day period from February16–24, 2020. Subsequently, a series of meetings were held with national institutions responsible for management, implementation, and evaluation of the response, particularly the National Health Commission and the China Centre for Disease Control and Prevention (China-CDC) to have first-hand information on a wide range of scenarios to understand the process of transmission and intensification of the pandemic in neighboring provenances and states. In this context, visits were conducted to the Beijing Municipality and the provinces of Sichuan (Chengdu), Guangdong (Guangzhon Shenzhen), and Hubei (Wuhan). The members of the Joint Mission visited health clinics, hospitals, temporary healthcare shelters, transportation hubs (air, rail, road), outdoor markets, COVID-19-designated hospitals, warehouses (personal protection equipment stock), and local centers for disease control (provincial and prefecture) to discuss and provide consultation to municipal mayors and their emergency operation teams, senior scientists, frontline clinical, public health, and community workers, and community neighborhood administrators. The findings in this report include discussion on control and prevention measures with national and local experts and response teams, and observations made and insights gained during site visits. The final report of the Joint Mission was submitted on February 28, 2020.

5.5.3 Women's leadership

The role of women in the novel coronavirus (COVID-19) pandemic is vital in mitigating health risk issues, within a framework of time and immediately available finance. Globally, women play an imperative role in discharging unpaid healthcare work, under high risk of financial insecurity, violence, exploitation, abuse, or harassment during times of crisis and quarantine. Women are one of the most vulnerable groups, being in the severe grip of COVID-19 pandemic, which may continue for an unpredictable period around the world.

In dealing with COVID-19, the role of gender in morbidity and mortality is critical. Men are noticed to be more prone to COVID-19 in terms of morbidity and mortality [5,106,107].

5.5.3.1 Compounded caregiving burden on women during the COVID-19 pandemic

As COVID-19 has spread globally, the impact of the pandemic on women has been intensifying severely. Women are serving at the forefront against the pandemic as they make up about 70% of the healthcare workforce, and thus expose themselves to greater risk of infection, and serve in leadership and decision-making roles in the healthcare sector. Many frontline women workers discharge double care responsibilities, being at their workplace for longer hours due to the pandemic crisis and carrying out unpaid healthcare work for children and elderly family members due to lockdown or curfew at home. While risky health service is equally important for both women and men, women tend to be overrepresented in certain types of vulnerable jobs without any protection measures, as compared to men. Due to heavy workloads, they get less time to take care of themselves. Many women who provide care in others' homes do not have employment contracts and work for uncertain periods with high levels of mental tension. Women and girls' unpaid care work is unavoidable, but an undervalued contributor to economies. Besides daily household work (such as cooking, cleaning, and taking care of children's education), their involvement in COVID-19-contaminated or pandemic-prone workplaces is crucial (Fig. 5.12).

5.5.3.2 Social protection and gendered risk

According to the OECD Development Centre's Social Institution and Gender Index (SIGI), globally, women are involved in significantly more care work than men. However, unpaid care provided by women and girls has many issues, like exposure to infection and psychosocial effects from providing care to an infected relative.

Currently most countries are using social protection measures to resolve some of the socioeconomic costs of both the pandemic and the containment measures, particularly on vulnerable groups, focusing on gender inequality. These initiatives include social assistance (e.g., family or child grants) and social insurance (e.g., unemployment insurance). In addition, provision for supporting low-income or vulnerable workers, paid sick leave, and waivers on rent and utilities payments have also been applied.

However, unpaid care responsibilities are being compounded due to closing of schools, offices, and other workplaces, and a global economic crisis looms. This problem may jeopardize action on gender equality and social exclusion. Care must be taken of the longer-term impacts of COVID-19 on gender and multidimensional poverty, keeping in view the social protection responses that do not address the fundamental range of gender inequality, including unpaid care and responsibilities. As the COVID-19 pandemic deepens women's inequality, it is necessary to minimize gender gap and fight against the COVID-19 pandemic with human spirit.

FIG. 5.12

Perform of women's unpaid work in (A) home, (B) hospital, (C) cooking at home, and (D) agriculture field.

5.5.3.3 COVID-19 impact on women and girls' unpaid care

The abrupt outbreak of the coronavirus pandemic has resulted in the closure of many services including schools, basic healthcare, and daycare centers, and shifting their provision on to women and girls within households. This has resulted in women feeling helpless, due to shortage of money to take proper care of family members. Even before the pandemic, globally, women and girls carried out on average three times the amount of unpaid care and domestic work of men and boys. These responsibilities will only increase with the new health and hygiene requirements, such as hand-washing and taking care of sick family members. Curfews and self-quarantine measures are likely to make these tasks even more challenging. Care burdens will manifest differently based on women and girls' ages and stages in life.

5.6 How to reduce women's domestic workload

Women in the workforce are the central pillar in resolving poverty reduction, food security, microfinance, and family well-being because they are responsible for both production and reproduction. Rural women in developing countries work for more

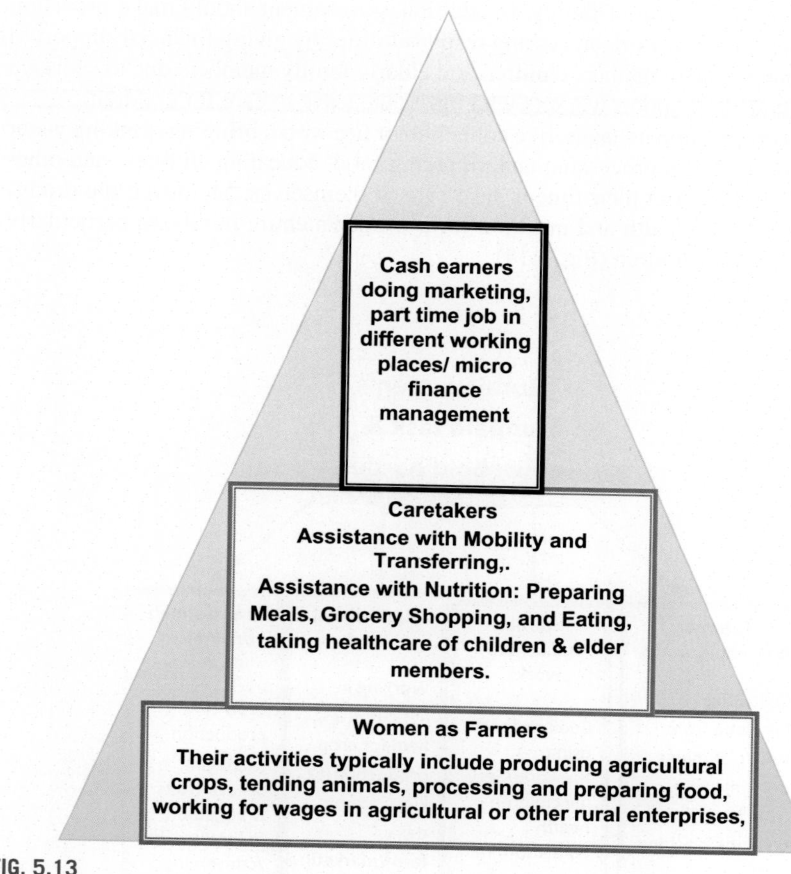

FIG. 5.13

Triple roles of rural women in overall development of communities.

time than men, by comparison. This is mainly due to multiple roles of rural women as farmers, income generating activities, and family care taker (Fig. 5.13). The multiple roles of women build up additional pressure on rural women's time. Women's heavy workload limits their time to spare for themselves and participation in other activities of their choice. So, in order to minimize the un paid labor of rural women, innovative technology to be developed in the field of nonconventional energy, drinking water supply system and farm related techniques.

Women are central to overcoming rural poverty. They play a critical role in poverty reduction and food security.

During corona like pandemic, the government of each state or authority in each province should find out some alternate means of earning or provide financial support or subsidiary to small retail business owners in order to secure their normal finance expenses for the survival of family. In addition, government should make provision to help parents in both work and caring responsibilities by giving financial support to those people unable to manage children and elderly family members due to shortage of money, and also support workers who must take leave to care for children.

Rural women's unpaid tasks, like collection of fire wood, bringing drinking water from a distance place, processing and preparing food, educating children, and other household work restrict their time to take care of themselves. Moreover, the drudgery can cause poor health and nutrition for a women's entire family, in particularly infants and young children (Fig. 5.14).

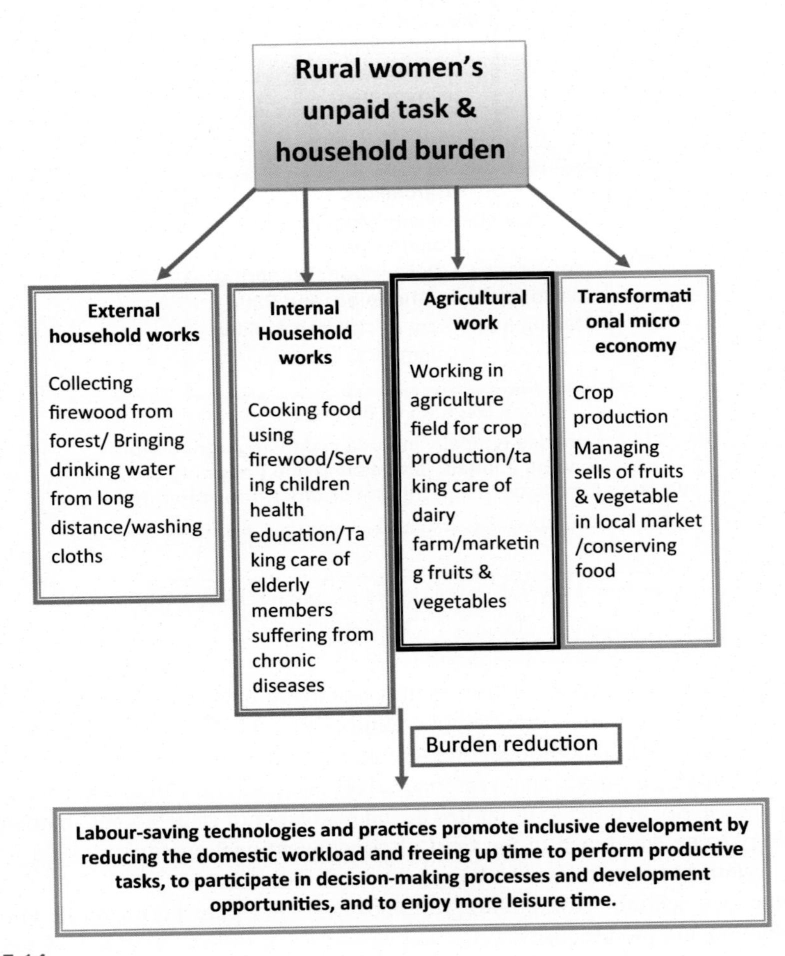

FIG. 5.14

Rural women's unpaid tasks and household burden, agricultural work, and transformational microeconomy; and possible methods for reduction.

This overburden of unpaid tasks not only impacts rural women but also affects smallholder farmers in developing agricultural productivity and achieving food and nutrition security.

Laborsaving technologies and their practical implementation can promote inclusive development by minimizing unpaid domestic tasks and giving opportunities to rural women to participate in various activities in different workplaces and decision-making bodies, and providing time to empower themselves to be eligible for developing small start-ups by using locally available resources. Women are the principal beneficiaries of reducing domestic workload, but men also benefit, depending on the extent to which they perform these tasks.

So, the Food and Agriculture Organization promotes innovative technology that can be adapted to different agro-ecological zones and climates, and which can be helpful in saving food waste and providing access to markets. Persistently high levels of rural poverty the rural people lead life under financial stress condition. So, FAO encourages and supports the rural people to develop innovative technology, based on local resources. A single labor-saving technology may not be workable for result-oriented achievement unless an integrated labor-saving system is adopted (Fig. 5.15).

Currently, some developing countries are assisting rural women to overcome additional burdens (Fig. 5.16) by adopting laborsaving technologies like improved stoves, biogas-based stoves, rainwater harvesting, and intermediate transport devices to overcome the problems of firewood collection and use for cooking, bringing drinking water from distant sources, and nonavailability of local transport systems, etc.

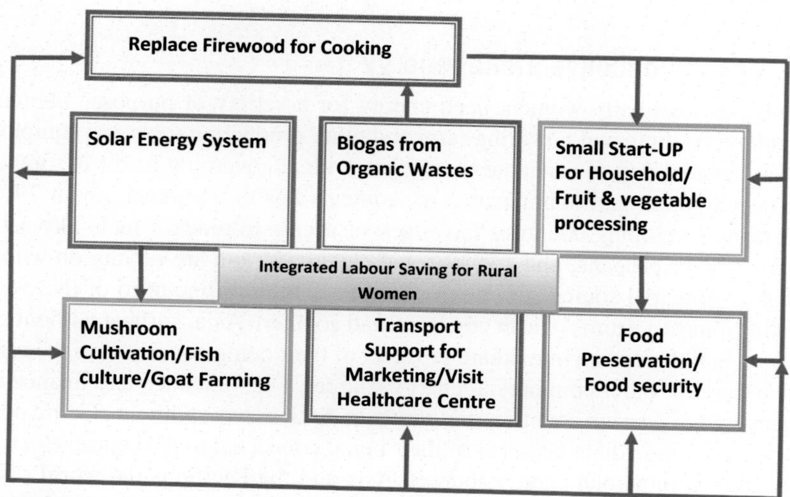

FIG. 5.15

LS integrated labor saving model for rural women.

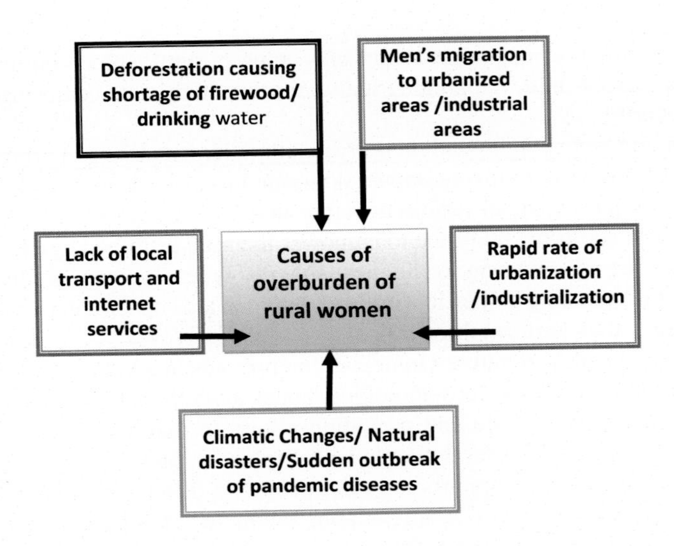

FIG. 5.16

Various causes of burden overload on rural women.

There are numerous laborsaving programs and projects operating in the rural areas of developing countries, but rural women are still overburdened. Rapid rates of urbanization and deforestation have forced women to walk for long distances to collect firewood and water. Migration of men for better earning opportunities has compounded the burden of agricultural work on women. Sudden outbreaks of pandemic diseases like COVID-19, and HIV and AIDS cause tremendous shortages of agriculture labors.

5.6.1 Use of nonconventional energy

Rural people, especially women, need energy for a variety of purposes like cooking, lighting, heating, and powering farm and other production tools and equipment. Therefore, availability of adequate energy is a critical governing factor in rural economic development and maintenance of women's health. However, about 79% of people from developing countries have lack of access to modern fuels like natural gas, kerosene, or propane, and conventional electricity, and are relying on wood or charcoal as principal sources of energy. About 1.5 billion (one-third of the world's population), mostly from African countries and southern Asia, survive without electric power, and there is no immediate prospect of their being connected to the central electricity grid or other commercial energy sources. Due to inadequate finance budget, many part of African countries and southern Asia, survive without electric power, and there is no immediate prospect of their being connected to the central electricity grid or other commercial energy sources. Asia and the Pacific is the world's faster growing region, and requires increasing energy supplies to fuel its rapid pace of economic expansion, including rural areas of developing countries. They account for 52% of the world's population, and need about 39% of global primary energy supply.

It is necessary, in rural areas of developing countries, to shift over from inefficient, traditional domestic burning of biomass, organic waste, and animal manure to biogas. Therefore, great efforts are being made to install small biogas digesters in rural areas, mainly in Asia, South America, and Africa (Fig. 5.17). In East Asia, considerable progress has been made in laborsaving technologies, especially for rural women. The West Guangxi Poverty-Alleviation Project in China (2002–2008) supported the Government's biogas program, including the provision of 22,500 bio-digesters to

FIG. 5.17

Showing shifting from traditional type of fire wood burning to domestic type biogas device.

about 30,000 households. Currently, enhancing bio-natural gas yield is one of main objectives of China's energy sector, which targets $300 \times 10^{27}\,m^3$ of biogas production in the market.

India is also a primary country where biomass power generation capacity has been gaining acceptability, especially in rural areas. By the end of 2017, the grid-connected biomass power generation capacity was 8.4 GW. It is expected that by the end of 2022, India will achieve the target of producing 10 GW. Globally, the share of biomass in total renewable energy power is 14%, as per the latest report from the World Bioenergy Association.

5.6.2 Water for drinking and household use

In most rural localities, water is not supplied to premises and needs to be collected. Rural women and girls are the only available unpaid workforce to collect water for drinking and household use from long distances, at the cost of their time and health (Fig. 5.18).

The main target of the UN's Sustainable Development Goals (SDGs) Goal 6 is universal and equitable access to safe and affordable drinking water by 2030. It also requires that everyone is within 30 min walk of some alternate means to get water in rural localities. However, the UN estimates are that in Sub-Saharan Africa, one roundtrip to collect water is 33 min on average in rural areas and 25 min in urban areas. In Asia, the numbers are 21 and 19 min, respectively. Furthermore, for particular countries, the figures may be higher. A single trip takes longer than an hour in Mauritania, Somalia, Tunisia, and Yemen. When water is not piped to the home, the burden of fetching it falls disproportionately on women and children, especially girls. On the basis of the UN's survey report from 24 Sub-Saharan countries, the water collection time is more than 30 min. It was estimated that 3.36 million children and 13.54

FIG. 5.18

Village people fetching water from distance.

million adult females were responsible for water collection. Collection of water from a distance not only causes problem to women and children, but the whole family also suffer. Sometimes, stored water gets contaminated and is responsible for diarrheal disease, which is the fourth leading cause of death among children under 5 years, and a leading cause of chronic malnutrition or stunting, which affects 159 million children worldwide. More than 300,000 children under 5 die annually from diarrheal diseases due to poor sanitation, poor hygiene, or unsafe drinking water—over 800 per day.

The Lower Usuthu Smallholder Irrigation Project (2003–13) in Swaziland trained women to construct water-harvesting tanks to improve access to water and promote income-generating activities. Most of the families with tanks now grow vegetables, both for home use and sale (Fig. 5.19).

FIG. 5.19

(A) Swaziland trained women to construct water-harvesting tanks to improve access to water and promote income-generating activities and (B) demonstration of tank growing vegetable for home use and sale.

FIG. 5.20

Borehole for water for use in household and drinking purpose in Nigeria.

The division of unpaid work between rural women and men in West and Central Africa is highly inequitable. Women work long hours every day carrying out domestic and agricultural work. In order to overcome the burden of water carrying from long distances, the Project for the Promotion of Local Initiatives for Development in Aguie (2005–13) in Nigeria built 20 village wells and 15 boreholes. This resulted in easy access to drinking water and saved time for rural women and girls walking distance for drinking water collection (Fig. 5.20).

5.6.3 Transport

Provision of local transport systems in rural areas of developing countries is extremely poor. Women have to spend a lot of time walking to local markets to sell produce. In South Asia, rural women typically work longer hours as compared to men, including domestic activities and unpaid agriculture tasks. In order to meet the requirement of household necessity and to visit primary health centers, women have to spend a lot of time in reaching their destination by walking. However, currently, a lot of countries are exploring the possibility of dealing with the transport problem by constructing roads to create links between urban and rural areas. The Agriculture, Marketing and Enterprise Promotion Programme (2005–12) in Bhutan constructed and rehabilitated 460 km of feeder road linked to rural areas. Haryana, a small agriculture-based state in India, has a network of 34 National Highways (NH) with a total length of 2484 km. The government has been working on the process of developing feeder roads between national highways and nearby rural areas to provide opportunity to farmers to market their agricultural products, including vegetables and fruits, to urban areas. Linking rural areas by a feeder road system provides rural communities with access to markets, easing the transport of goods, and enables more shops to open in rural areas (Fig. 5.21).

FIG. 5.21

Roadside markets for promoting rural women for selling fruits and vegetables.

Availability of local shops allows women to spend a mere few minutes buying household items rather than a full day traveling to the main market for the same goods. The time saved is used for vegetable production, an important source of income and nutrition.

5.6.4 Mix response

Migration of men from rural areas for better earning potential has been a serious issue around the world. Men's migration creates a compounded burden of unpaid work for rural women. Male workers from the neighboring rural areas of Central America and Mexico migrate to urban area for job opportunities and better wages, which results in additional unpaid workload on women. The Rural Development and Modernization

Project for Eastern Region Development and Modernization Project for the Eastern Region (2005–13) in El Salvador were launched for social wellbeing. The former project aimed to improve the standard of living and earning of the residents of 33 municipalities in eastern El Salvador. Most of the farmers in this area are beneficiaries of the project for the production of coffee and sugarcane harvests. The project was also helpful in promoting marketing to neighboring urban areas for better profit. The latter project was aimed at reducing the domestic workload of women and facilitating their involvement in productive activities. Rural women were provided with improved stoves, mills, community kitchens, household water cisterns, agroforestry parcels to decrease the need to collect firewood, and child day-care centers. All these provisions resulted in women saving time and allowed them to take part in other activities.

The Rural Income Diversification Project (RIDP) in Tuyen Quang Province (2002–10) in Vietnam, was a campaigning training program to bring awareness on gender equality, division of labor, the prevention of domestic violence, and women's participation in decision-making. Well-design protocols on construction of safe water supply systems, latrines, and children nurseries are provided to rural women and girls, and provision of free supplies of threshing machines and scholarships for schoolchildren from poor households were also included in the training program. The primary objective was to improve the socioeconomic status of 49,000 poor households living in upland areas, especially ethnic minorities and women. The critical change the project intended to introduce was the enhanced linkage between small farmers and markets, as well as support to crop and livestock production, building access roads to markets, and providing vocational training for rural youth. The project also endeavored to promote microenterprises, with a view to moving local production up the value chain.

Rural women's unpaid care workload (both domestic and agriculture) is a long-standing chronic issue , and one of the main reasons for rural poverty and poor health condition. Laborsaving technologies and practices challenge the present generation to develop innovative simple technologies that are helpful to rural women in saving time from their unpaid care workload to look after themselves and their children, even under unfavorable conditions. Therefore, many international agencies, such as the International Fund for Agricultural Development (IFAD), UN (thru the Sustainable Development Goals), FAD, ILO, and UNICEF, have objectives to eradicate poverty, which is the root cause of rural women's unpaid care workload, poor health, hunger, and food insecurity. Keeping in view the global economic, environmental, and socio-cultural changes, maximum efforts are under process at various voluntary organization and government bodies to bring alternate innovative low cost technology to minimize the issues related to unpaid work load on rural women and their quality of life. It is also important to consider simple innovative technologies instead of designing highly expensive technology based on sophisticated computerized devices. Innovative laborsaving technology should preferably be based on locally available low cost base materials, which can be easily acceptable to rural people, without any heavy out-of-pocket expenditure. It should also be taken into account that technology development needs to incorporate women's perspective and the development of women's innovative and technological capabilities to enable them to better solve their own problems in response to the rapid changes taking place around them.

References

[1] Farha L. Committee on the elimination of discrimination against women. In: Langford M, editor. Social rights jurisprudence: emerging trends in international and comparative law. Cambridge University Press; 2008. p. 560–1.

[2] Building on achievements: women's human rights five years after Beijing. May 2000, paras. 9–22. www.ohchr.org.

[3] United Nations Human Right (2012) Born Free and Equal: Sexual Orientation and Gender Identity in International Human Rights Law (HR/PUB/12/06).

[4] Office of the High Commissioner for Human Rights OHCHR. Women facing multiple forms of discrimination., 2009, www.un.org/en/durbanreview2009/pdf/InfoNote_07_.

[5] Zhu N, Zhang D, Wang W, Li X, Yang B, Song J, et al. A novel coronavirus from patients with pneumonia in China, 2019. N Engl J Med 2020;382:727–33. https://doi.org/10.1056/NEJMoa2001017.

[6] National Health Commission of PRC. Daily briefing on novel coronavirus cases in China., 2020, http://ennhcgovcn/2020-02/23/c_76779htm.

[7] Brooks SK, Webster RK, Smith LE, Woodland L, Wessely S, Greenberg N, Rubin J. The psychology impact of quarantine and how to reduce it. Lancet 2020;395:912–20. https://doi.org/10.1016/S0140-6736(20)30460-8.

[8] Peterman P, O'Donnell T, Shah O-P, van Gelder. Pandemics and violence against women and children. CGD Working Paper 528, Washington, DC: Center for Global Development; 2020. https://www.cgdev.org/publication/pandemics-and-violence-against-women-and-children.

[9] Worldometer Coronavirus., 2020, https://www.worldometers.info/coronavirus/country/turkey/.

[10] Chen B, Liang H, Yuan X, Hu Y, Xu M, Zhao Y, Zhang B, Tian F, Zhu X. Roles of meteorological conditions in COVID-19 transmission on a worldwide scale. MedRxiv 2020. https://doi.org/10.1101/2020.03.16.20037168.

[11] Urban LYKRC. COVID-19 pandemic: impacts on the air quality during the partial lockdown in São Paulo state. Brazil Sci Total Environ 2020;730. https://doi.org/10.1016/j.scitotenv.2020.139087.

[12] Saadat S, Rawtani D, Hussain CM. Environmental perspective of COVID-19. Sci Total Environ 2020;728. https://doi.org/10.1016/j.scitotenv.2020.138870.138870.

[13] Blustein DL, Duffy R, Ferreira JA, Cohen-Scali V, Cinamon RG, Allan BA. Unemployment in the time of COVID-19: a research agenda. Elsevier; 2020.

[14] Jain R, Budlender J, Zizzamia R, Bassier I. The labor market and poverty impacts of Covid-19 in South Africa; 2020.

[15] Statistics South Africa. Quarterly Labour Force Survey Quarter 2. Statistical release P0211. Pretoria: Statistics South Africa; 2020.

[16] Casale D, Posel D. Gender inequality and the COVID-19 crisis: evidence from a large national survey during South Africa's lockdown. Res Soc Strat Mobil 2020;, 100569.

[17] Wills G, Patel L, Van der Berg S, Mpeta B. Household resource flows and food poverty during South Africa's lockdown: short-term policy implications for three channels of social protection. Working Paper Series NIDS-CRAM Wave 1; 2020.

[18] Oyenubi A, Kollamparambil U. COVID-19 and depressive symptoms in South Africa; 2020. Report No.: 10.

[19] Vindegaard N, Benros ME. COVID-19 pandemic and mental health consequences: Systematic review of the current evidence. Brain Behav Immun 2020. 32485289.

[20] Xiong J, Lipsitz O, Nasri F, Lui LM, Gill H, Phan L, et al. Impact of COVID-19 pandemic on mental health in the general population: a systematic review. J Affect Disord 2020.
[21] Banks J, Xu X. The mental health effects of the first two months of lockdown during the COVID-19 pandemic in the UK. Fiscal Stud 2020;41:685–708.
[22] Proto E, Quintana-Domeque C. COVID-19 and mental health deterioration by ethnicity and gender in the UK. PLoS One 2021;16, e0244419. 33406085.
[23] Graetz B. Health consequences of employment and unemployment: longitudinal evidence for young men and women. Soc Sci Med 1993;36:715–24. 8480216.
[24] Murphy GC, Athanasou JA. The effect of unemployment on mental health. J Occup Org Psychol 1999;72:83–99.
[25] Burgard SA, Brand JE, House JS. Toward a better estimation of the effect of job loss on health. J Health Soc Behav 2007;48:369–84. 18198685.
[26] Paul KI, Moser K. Unemployment impairs mental health: Meta-analyses. J Voc Behav 2009;74:264–82.
[27] Tomlinson M, Grimsrud AT, Stein DJ, Williams DR, Myer L. The epidemiology of major depression in South Africa: results from the south African stress and health study. S Afr Med J 2009;99. 19588800.
[28] Evans-Lacko S, Knapp M, McCrone P, Thornicroft G, Mojtabai R. The mental health consequences of the recession: economic hardship and employment of people with mental health problems in 27 European countries. PLoS One 2013;8, e69792. 23922801.
[29] Nwosu CO. The relationship between employment and mental and physical health in South Africa. Dev South Afr 2018;35:145–62.
[30] IMF, add refs https://blogs.imf.org/2020/04/06/an-early-view-of-the-economic-impact-of-the-pandemic-in-5-charts/.
[31] https://www.forbes.com/sites/miltonezrati/2020/03/18/heading-off-the-covid-19-recession/#651eba9a28e6.
[32] UN Women calculations based on Oxford COVID-19 government response tracker. https://www.bsg.ox.ac.uk/research/research-projects/oxford-covid-19-government-response-tracker.
[33] http://www.ugogentilini.net/wp-content/uploads/2020/04/Country-social-protection-COVID-responses_April3-1.pdf.
[34] http://pubmed.ncbi.nlm.nih.gov. Out-of-pocket spending for contraceptives in Latin America. UNFPA, Latin America and Caribbean Regional Office, March 2020; 2020.
[35] International Labour Organisation. Care work and care jobs for the future of decent work; 2018.
[36] ILO. ILO monitor: COVID-19 and the world of work. 5th ed., 30 June 2020, Geneva: ILO; 2020.
[37] UN Women. Progress of the world's women 2015–2016: transforming economies, realizing rights. New York: UN Women; 2015.
[38] Giannini S, Jenkins R, Saavedra J. Reopening schools: when, where and how? 2020, https://en.unesco.org/news/reopening-schools-when-where-and-how.
[39] UN. The impact of COVID-19 on women. Policy brief. New York: United Nations; 2020.
[40] World Health Organization (WHO). Delivered by women, led by men: a gender and equity analysis of the global health and social workforce. Human resources for health observer—issue no. 24. Geneva: WHO; 2019.
[41] Gross CP, Essien UR, Pasha S, Gross JR, Wang SY, Nunez-Smith M. Racial and ethnic disparities in population level Covid-19 mortality. MedRxiv 2020. https://doi.org/10.1101/2020.05.07.20094250.

[42] Razaq A, Harrison D, Karunanithi S, Barr B, Asaria M, Khunti K. Hidden in plain sight: BAME COVID-19 deaths—what do we know? Rapid Data Evid Rev 2020. https://www.cebm.net/wp-content/uploads/2020/05/BAME-COVID-Rapid-Data-Evidence-Review-Final-Hidden-in-Plain-Sight-compressed.pdf.
[43] Folbre N. Developing care: recent research on the care economy and economic development; 2018.
[44] United Nations. Leaving no one behind: a call to action on gender equality and women's economic empowerment; 2016.
[45] World Bank. Gender equality and development; 2012.
[46] Addati L, Cattaneo U, Esquivel V, Valarino I. Care work and care jobs for the future of decent work. Geneva: International Labour Organization; 2018.
[47] Fletcher E, Pande R, Moore CM. Women and work in India: descriptive evidence and a review of potential policies. HKS working paper no. RWP18-004, Cambridge: Center for International Development Harvard University; 2017.
[48] Mehrotra S, Parida J, Sinha S, Gandhi A. Explaining employment trends in the Indian economy: 1993–94 to 2011–12. Econ Pol Wkly 2014;49(32):49–57.
[49] Crow B, McPike J. How the drudgery of getting water shapes women's lives in low-income urban communities. Gend Technol Dev 2009;13(1):43–68.
[50] Patel SJ, Patel MD, Patel JH, Patel AS, Gelani RN. Role of women gender in livestock sector: a review. J Livest Sci 2016;7:92.
[51] Hirway I. Unpaid work and the economy: linkages and their implications. Indian J Labor Econ 2015;58(1):1–21.
[52] Addati L, Cattaneo U, Esquivel V, Valarino I. Care work and care jobs for the future of decent work. Geneva: International Labor Organization; 2018.
[53] Women's economic empowerment in the changing world of work, Report of the Secretary-General, E/CN.6/2017/3, 2016.
[54] Mincer J. Labor force participation of married women: a study of labor supply. In: Aspects of labor economics. New Jersey: Princeton University Press; 1962. p. 63–105.
[55] Becker GS. A theory of the allocation of time. Econ J 1965;75(229):493–517.
[56] Benston M. The political economy of women's liberation. Mon Rev 1969;21(4):13–27.
[57] Dalla Costa M, James S. The power of women and the subversion of the community. IMF working paper WP/15/55, Washington: Asia and Pacific Department; 1975.
[58] Harrison J. The political economy of housework. Bull Conf Soc Econ 1973;3(3):35–52.
[59] Gardiner J, Himmelweit S, Mackintosh M. Women's domestic labor. New Left Rev 1975;89(1):47–58.
[60] Folbre N, Yoon J. Economic development and time devoted to direct unpaid care activities. Geneva: UNRISD Flagship Report on Poverty. United Nations Research Institute for Social Development; 2008.
[61] Marlowe F. Hunting and gathering: the human sexual division of foraging labor. Cross-Cultural Res 2007;41(2):170–95.
[62] Cott NF. The bonds of womanhood: woman's sphere in new England 1780–1835. Bristol: Falling Wall Press Bristol; 1997.
[63] Hartmann H. Capitalism, patriarchy, and job segregation by sex. Signs (Chic) 1976;1(3/2):137–69.
[64] Dubbert JL. A man's place: masculinity in transition. Prentice Hall: The University of Virginia; 1979.
[65] Matthaei J. Healing ourselves, healing our economy: paid work, unpaid work, and the next stage of feminist economic transformation. Rev Radic Polit Econ 2001;33(4):461–94.

[66] Kessler-Harris A. Equal employment opportunity commission v. Sears, roebuck and company: a personal account. Fem Econ 1987;25(1):46–69.
[67] Williams J. Unbending gender: why family and work conflict and what to do about it. Oxford: Oxford University Press; 2001.
[68] Amott TL, Matthaei JA. Race, gender, and work: a multi-cultural economic history of women in the United States. South End Press; 1996.
[69] Elson D. Recognize, reduce, and redistribute unpaid care work: how to close the gender gap. New Labor Forum 2017;26(2):52–61.
[70] Hirway I. Unpaid work and economy: gender, poverty and millennium development goals. Integrating unpaid work into development policy working paper no. 838. Hudson; 2005.
[71] Kabeer N. Women's economic empowerment and inclusive growth: labor markets and enterprise development. Int Dev Res Cent 2012;44(10):1–70.
[72] Maloney WF. Informality revisited. World Dev 2004;32(7):1159–78.
[73] Aguirre D, Hoteit L, Rupp C, Sabbagh K. Empowering the third billion: women and the world of work in 2012. New York: Booz and Company; 2012.
[74] Grimshaw D, Rubery J. The motherhood pay gap. Working paper no. 1/2015, Geneva: International Labor Organization; 2015.
[75] Neff DF, Sen K, Kling V. The puzzling decline in rural women's labor force participation in India: a reexamination. Working paper no. 196, Hamburg: GIGA German Institute of Global and Area Studies; 2012.
[76] Das MS, Jain-Chandra S, Kochhar MK, Kumar N. Women workers in India: why so few among so many? Washington: International Monetary Fund; 2015.
[77] Budlender D. Why should we care about unpaid care work? Southern Africa Regional Office; 2004.
[78] Razavi S. The political and social economy of care in a development context: conceptual issues, research questions and policy options (3). Switzerland: United Nations Research Institute for Social Development Geneva; 2007.
[79] Mitra S. Women and unpaid work in India: a macroeconomic overview. IWWAGE 7 March 2019.
[80] MacLeavy J. Care work, gender inequality and technological advancement in the age of COVID-19. Gend Work Organ January 2021;28(1).
[81] Kashen J. How COVID-19 relief for the care economy fell short in 2020. The Century Foundation; 27 January 2021.
[82] Diallo B, Qayum S, Staab S. COVID-19 and the care economy: immediate action and structural transformation for a gender-responsive recovery. UN Women; 2020.
[83] Turquet L, Koissy-Kpein S. Covid-19 and gender: what do we know; what do we need to know? UN Women; 2020.
[84] Prewitt L. Making care count. Care Work Econ 24 May 2021.
[85] UNESCO. Take a survey: COVID-19 and early childhood education workforce. https://en.unesco.org/news/take-survey-covid-19-and-early-childhood-education-workforce; 2020.
[86] https://qz.com/india/1826683/indias-approach-to-fighting-coronavirus-lacks-a-gender-lens/.
[87] https://www.scmp.com/economy/china-economy/article/3078519/world-risk-second-great-depression-due-coronavirus-says.
[88] Boniol M, McIsaac M, Xu L, Wuliji T, Diallo K, Campbell J. Gender equity in the health workforce: analysis of 104 countries: Working paper 1. Geneva: World Health Organization; 2019.
[89] Addressing the emerging impacts of the COVID-19 pandemic on migrant women in Asia and the Pacific for a gender-responsive recovery. UN Women Guidance for Action

Brief. Bangkok May. 2020. https://asiapacific.unwomen.org/en/digital-library/publications/2020/04/guidance-for-action-addressing-the-emerging-impacts-of-the-covid-19-pandemicnder-responsive-recovery.

[90] Wenham, Smith, Morgan. Covid 19: the gendered impacts of the outbreak. Lancet 2020. 6.3.2020. https://www.thelancet.com/journals/lancet/article/PIIS0140-6736(20)30526-2/fulltext.

[91] Goyal, Brar. After four test positive, Punjab ASHA workers ask: 'Why should we put our lives in danger?'. Indian Exp 2020. 04.05.202. https://indianexpress.com/article/india/after-four-test-positive-asha-workers-ask-why-should-we-put-our-lives-in-danger-6392288/.

[92] Kabeer N. Gender equality, economic growth, and women's agency: the 'endless variety' and 'monotonous similarity' of patriarchal constraints. Fem Econ 2016;22(1):295–321.

[93] International Labour Organization. Care work and care jobs for the future of decent work. Geneva: ILO; 2018.

[94] Charmes J. The unpaid care work and the labour market. An analysis of time use data based on the latest world compilation of time-use surveys. Working paper, Geneva: ILO; 2019.

[95] Dugarova E, Gülasan N. Global trends: challenges and opportunities in the implementation of the sustainable development goals. Joint report by the United Nations Development Programme and the United Nations Research Institute for Social Development. New York/Geneva: UNDP/UNRISD; 2017.

[96] UNFPA and HelpAge International. Ageing in the 21st century: a celebration and a challenge. New York/London: HelpAge International/UNFPA; 2012.

[97] Kan M, Sullivan O, Gershuny J. Gender convergence in domestic work: discerning the effects of interactional and institutional barriers from large-scale data. Sociology 2011;45(2):234–51.

[98] Dugarova E. Gender equality as an accelerator for achieving the sustainable development goals. New York: UN; 2018.

[99] Esquivel V, Kaufmann A. Innovations in care. In: New concepts, new actors, new policies. Berlin: Friedrich Ebert Stiftung; 2017.

[100] Heintz J. Why macroeconomic policy matters for gender equality. New York 2015.

[101] Hess C, Hegewisch A. The future of care work: improving the quality of America's fastest-growing jobs. IWPR report #C486, Washington, DC: Institute for Women's Policy Research; 2019.

[102] Hess C, Milli J, Hayes J, Hegewisch A. The status of women in the states. Washington, DC: Institute for Women's Policy Research; 2015. IWPR report#C400.

[103] Hess C, Ahmed T, Hayes J. Providing unpaid household and care work in the United States: uncovering inequality. Institute for Women's Policy Research: Briefing paper; 2020.

[104] Dugarova E, Women's Budget Group. Crises collide: women and Covid-19. Examining gender and other equality issues during the coronavirus outbreak. Women's Budget Group; 2020. http://un.orgs>Durragove.

[105] Benard S, Paik I, Correll S. Cognitive bias and the motherhood penalty. Hastings Law J 2008;59(6):1359. ISSN 0017-8322.

[106] Jin JM, Bai P, He W, Wu F, Liu XF, Han DM, et al. Gender differences in patients with COVID-19: focus on severity and mortality. medRxiv [Preprint] 2020. https://doi.org/10.1101/2020.02.23.20026864.

[107] Li Q, Guan X, Wu P, Wang X, Zhou L, Tong Y, et al. Early transmission dynamics in Wuhan, China, of novel coronavirus-infected pneumonia. N Engl J Med 2020;382:1199–207. https://doi.org/10.1056/NEJMoa2001316.

Index

H

I

Printed in the United States
by Baker & Taylor Publisher Services